Questions and Solutions for Chemistry

Questions

John Sadler and Mike Taylor

Edward Arnold

© John Sadler and Mike Taylor 1986

First published in Great Britain 1986 by

Edward Arnold (Publishers) Ltd, 41 Bedford Square, London WC1B 3DQ

Edward Arnold (Australia) Pty Ltd, 80 Waverley Road,
Caulfield East, Victoria 3145, Australia

British Library Cataloguing in Publication Data

Sadler, John
  Questions and solutions for chemistry.
  Questions
  1. Chemistry—Examinations, questions, etc.
  I. Title   II. Taylor, Mike
  540'.76   QD42

  ISBN 0-7131-8410-8

All rights reserved. No part of this publication may be reproduced, stored in a retrieval system, or transmitted in any form or by any means, electronic, photocopying, recording, or otherwise, without the prior permission of Edward Arnold (Publishers) Ltd.

Text set in 10/11 Plantin
by Macmillan India Ltd., Bangalore 25.
Printed and bound Thomson Litho Ltd, East Kilbride, Scotland

Contents

Introduction

How to use the questions	1
How to use the solutions	2
Definitions	3
Abbreviations	4

1 Mixtures and compounds — 5

1.1 Chromatography	5
1.2 Mixtures and compounds	6
1.3 Solubility and separation of gases	6
1.4 Separation of mixtures	6
1.5 Separation of a mixture	6

2 Particles, diffusion and radioactivity — 7

2.1 Diffusion of gases 1	7
2.2 Diffusion of gases 2	8
2.3 Radioactive decay series	8
2.4 Radioactive elements	9
2.5 Nuclear reactions	9

3 Atomic structure and chemical bonds — 10

3.1 Atomic structure and bonding	10
3.2 Atomic structure 1	10
3.3 Atomic structure 2	11
3.4 Atomic number and bonding	12
3.5 Chemical bonding	12
3.6 Structure and bonding	13

4 Moles and equations — 14

4.1 Mole calculation	14
4.2 The mole/oxides of nitrogen	14
4.3 The mole/energy (enthalpy) change	15
4.4 The mole/potassium manganate(VII)	16
4.5 Mole calculation/precipitation	17

5 Electrolysis and electrical cells — 18

5.1 Conductors/non-conductors	18
5.2 Ions	19
5.3 Electrolysis of aqueous solutions	19
5.4 Electricity and fluorine	20
5.5 Electrolysis of copper(II) sulphate	21
5.6 Transition metals/electroplating	22
5.7 Redox/cells	22

6 Energy changes, speeds of reaction and reversible reactions — 23

- 6.1 Chemical reactions — 23
- 6.2 Speed of reaction 1 — 23
- 6.3 Speed of reaction 2 — 24
- 6.4 Speed of reaction 3 — 25
- 6.5 Speed of reaction 4 — 25
- 6.6 Catalysts — 27
- 6.7 Catalyst: chemical equilibrium — 27
- 6.8 Reversible reactions — 27

7 Salt preparations, strong and weak acids — 28

- 7.1 Salt preparation: copper(II) sulphate — 28
- 7.2 Salt preparation: calcium sulphate — 28
- 7.3 Ions — 29
- 7.4 Strong acid — 30
- 7.5 Strong and weak acids — 30
- 7.6 Organic acid — 31

8 The Periodic Table — 32

- 8.1 Periodic Table — 32
- 8.2 Periodic Table: Period 3 — 32
- 8.3 Periodic Table: Group VII — 33
- 8.4 Lithium — 33
- 8.5 Periodic Table: calculation — 34

9 The Halogens — 35

- 9.1 Chlorine — 35
- 9.2 Sodium chloride — 35
- 9.3 Fluorine — 36
- 9.4 Bromine — 36
- 9.5 Oxidation and reduction — 37

10 Nitrogen compounds — 38

- 10.1 Ammonia — 38
- 10.2 Nitric acid — 38
- 10.3 Manufacture of ammonia — 38
- 10.4 Fertilisers and nitric acid — 39
- 10.5 Nitrogen dioxide — 39

11 Carbon compounds, fuels and hard water — 40

- 11.1 Carbon dioxide — 40
- 11.2 Combustion of carbon — 40
- 11.3 Carbon monoxide — 41
- 11.4 Limestone/hardness of water — 41
- 11.5 Hardness of water 1 — 42
- 11.6 Hardness of water 2 — 42

12 Carbon chemistry — 43

- 12.1 Crude oil — 43
- 12.2 Alkanes — 43
- 12.3 Homologous series — 44
- 12.4 Methane and ethane — 44
- 12.5 Alkenes — 45
- 12.6 Organic structures — 45
- 12.7 Monomers and polymers — 46
- 12.8 Proteins — 46
- 12.9 Hydrocarbons/heats of combustion — 47

13 Metals and the reactivity series — 49

- 13.1 Reactivity series — 49
- 13.2 Transition metals — 49
- 13.3 Manufacture of iron — 50
- 13.4 Copper compounds — 50
- 13.5 Tests for metals — 51

14 Sulphur and sulphur compounds — 52

- 14.1 Sulphur — 52
- 14.2 Preparation of sulphur dioxide — 52
- 14.3 Sulphuric acid — 53
- 14.4 An industrial process — 53
- 14.5 Sulphates — 53

15 Air, oxygen, hydrogen and redox reactions — 54

- 15.1 Percentage of oxygen in the air — 54
- 15.2 Oxygen 1 — 55
- 15.3 Air — 55
- 15.4 Oxides — 56
- 15.5 Oxygen 2 — 56
- 15.6 Acidic and basic oxides — 57
- 15.7 Oxidation and reduction — 57
- 15.8 Hydrogen as a reducing agent — 58

16 Technological, social, economic and environmental chemistry — 59

- 16.1 Manufacturing processes — 59
- 16.2 Discovery of the elements — 60
- 16.3 Modern materials — 61
- 16.4 Disposal of household waste — 61
- 16.5 Limestone, lime and slaked lime — 62
- 16.6 Manufacture of sulphuric acid — 63
- 16.7 Cigarettes — 64

17 Analysis — 66

- 17.1 Tests for ions — 66
- 17.2 Analysis — 66

17.3 Analysis of a black compound	67
17.4 Analysis involving iron compounds	67
17.5 Analysis of a mineral	68

18 Various topics combined 70

18.1 Uses of various substances	70
18.2 Tests for various gases	70
18.3 Preparation of gases	70
18.4 Reactions of various sodium compounds	71
18.5 False/true statements	71
18.6 Properties of elements	72

Introduction

How to use the questions

The questions are arranged in sections, and numbered within each section, so that question 8.2 is the second question in section 8. The answers are numbered to correspond. The questions are of different length and they are not intended to represent actual examination questions.

There is a variety of styles within the questions, and different depths of treatment of the same topic may be expected in different questions. You will have to judge this by the number of marks allocated.

A For practice in answering examination questions

Choose the question you want to try (if your teacher has not chosen it for you). You might choose a question because you think you can answer it well, and want to see just how well; or because you don't think you know much about it, and want to find out what are the important bits to learn.
Don't look at the answers yet!!
Read the question slowly and carefully, and make sure that you understand what it means. Time spent doing this is never wasted, even in an examination. Look in particular to see if the words in the 'Definitions' are used in the question; examiners are very precise about what they mean and they expect you to be the same!

Look at the marks given for each part of the question so that you know where a lot is expected and where only a short answer will do. By and large, the more marks there are, the more you have to write or draw to get them. Always leave time for the short answers; those few marks might make all the difference in obtaining a high grade.

Write out your answers to the questions. If you can't do it all, do as much as you can. If there is a calculation, try it even if you can't finish it. There are marks for the method as well as for the answer. You should include equations, diagrams and graphs wherever these will help to illustrate your answer more clearly and logically.

Then look at the answers and see what you got right, what you missed out, and where you had the right idea but didn't express it properly.

B For quick revision or making sure you've learned something

Instead of writing out the answer you can just *think* what you would have written; then look at the answer and see if you had the scoring points. Read up those parts you didn't get right. Then think through the question again the *next* day.

How to use the solutions

No two people would write the same words in answer to a question. But there are always a certain number of important points which ought to be included; and because chemistry is an exact science, there are times when you have to say exactly what you mean. For example, it is of no use to say 'I would use sodium hydroxide' if you mean 'I would add a solution of sodium hydroxide'. One answer will get the mark, the other will not.

Of course, there are times when different answers are acceptable. 'Warm the solution' would be the same as 'Heat the solution gently'. 'Collect the gas by downward displacement of air' means the same as 'Collect the gas by upward delivery'. In a calculation, you may get the right answer by working in moles (which is always better) or in grams; correct working and a correct answer will get full marks whichever way you do it.

Note that if you make an error in the arithmetic you lose that mark only. If you then use your wrong figure in later work, there is no further penalty.

If, however, you make an error of chemical principle, e.g. saying that 1 g of a gas occupies 24 dm^3 at room temperature and pressure, you will lose more marks.

When you have written your answer, compare it with the answer in the book. If you have put down *exactly* what the answer says, then give yourself the mark. If you have not, you have to ask yourself 'does my answer mean exactly the same?' Once again, if the answer is yes you can give yourself the mark. If not, you have to decide whether you are close enough, and there your teacher will help.

Each * in the answer is one scoring point.

Notes

Fractions are given as x/y, not $\frac{x}{y}$.

It is better to calculate intermediate answers as you go through the calculation rather than carry through complex fractions.

In the diagrammatic representation of covalent molecules, show all the electron shells, unless the question clearly gives instruction otherwise.

Units are given either with the solidus, e.g. mol/dm^3, or with the negative index, e.g. mol dm^{-3}. Both these examples mean moles per cubic decimetre.

Any relative atomic masses (A_r) needed in the questions can be obtained from the information on the Periodic Table (page 73).

Definitions

'Calculate' means that you have to show the working and give the numerical answer with the correct units.

'Construct' (usually used for writing an equation) means use the information provided to work out the answer.

'Deduce' means some reasoning is required, such as reference to a law or principle.

'Define' means give the exact meaning of.

'Determine' means that you must use the information given to solve the problem.

'Describe' means say what you would actually do or see in enough detail that there is nothing for the reader to have to guess. Include any equations which would be helpful. You can also score the marks for 'describe' if you draw a diagram. If the question asks 'Describe, with the aid of a diagram . . .', you will lose marks if you do not draw a diagram.

'Equation' means a balanced molecular equation with all the formulae right. You should include state symbols.

'Estimate' (usually applied to numerical questions) means that you have to make an approximate judgement based on information given in the question.

'Explain' means say why; some reasoning or reference to the theory is needed. ('State and explain' is often used; you are expected to do both!)

'Find' is often used in place of 'calculate' and 'determine'.

'Full structural formula' means a formula in which every bond is shown:

$$H-\underset{\underset{H}{|}}{\overset{\overset{H}{|}}{C}}-\underset{\underset{H}{|}}{\overset{\overset{H}{|}}{C}}-O-H \text{ is a full structural formula}$$

but CH_3CH_2OH is not a full structural formula.

'Ionic equation' means an equation showing the ions which react together; the ion formulae must be correct, the ions must balance and the electrical charges must balance. Do not show spectator ions – i.e. those ions which do not take part in the reaction.

'Outline' means keep the answer short; only the essential details are required.

'Predict' means that you have to make a logical deduction from the information given in the question.

'Sketch' (usually for a graph) means that the shape and position of the graph on the axes need only be broadly right; but you might be expected to discuss such things as the curve passing through the origin or tending to a constant value.

'State' means give a factual answer, but don't bother to give supporting facts.

'Suggest' means give your own idea of the best answer. (There may be more than one possibility, and you choose the best; or you may be asked to apply your knowledge to a novel situation which you have never met – even one 'outside the syllabus'.)

Abbreviations

A_r	relative atomic mass
M_r	relative molecular mass
RFM	relative formula mass
r.t.p.	room temperature and pressure
dm^3	decimetre cube; $1\,dm^3 = 1000\,cm^3$
(s)	solid
(l)	liquid
(g)	gas
(aq)	aqueous – dissolved in water

Acknowledgements

We thank Mr Ken Yates, moderator of the 'O' level Chemistry and Science and Integrated Science papers, University of Cambridge Local Examinations Syndicate, for his careful reading of the manuscript and for his many helpful suggestions to improve the text.

We thank Alison Sadler for redrawing many of the illustrations.

Finally, we thank our families for their patience and encouragement in the preparation of this book, and for putting up with having a Chief Examiner in the house!

1 Mixtures and compounds

1.1 Chromatography

A mixture of dyes, M, was analysed by chromatography. The experiment was performed using two different solvents X and Z. The results were as shown below.

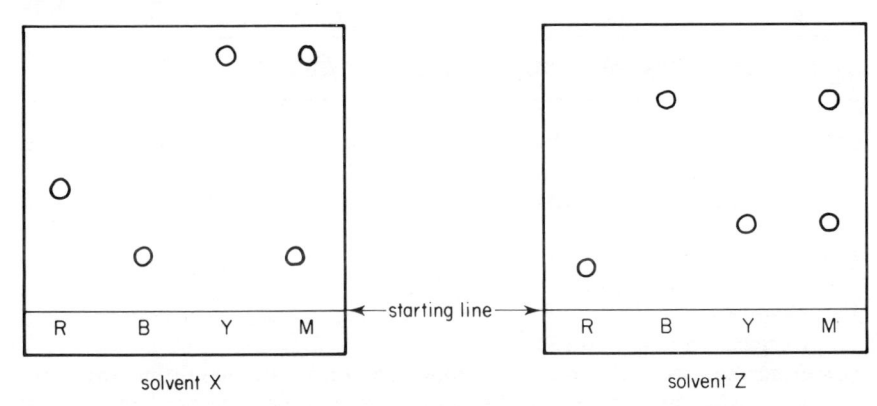

R = red   B = blue
Y = yellow   M = mixture

(a) Describe, with the aid of a diagram, how you would carry out one of these experiments, stating any precautions you would take. (6)
(b) Suggest two factors that contribute to the separation of the dyes. (2)
(c) (i) Which dyes are present in the mixture M? (1)
(ii) What is the colour of mixture M? (1)
(d) Draw a sketch to show the appearance of the chromatogram produced if mixture M was first developed in solvent X, then turned through 90 degrees and edge ST placed in solvent Z (see diagram). (3)

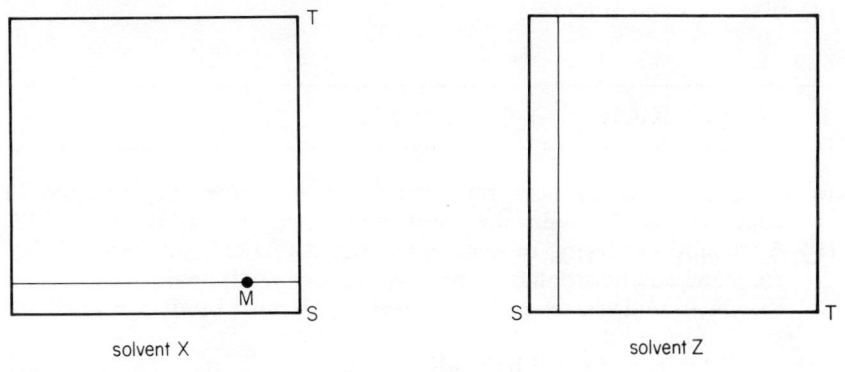

(e) State one industrial use of chromatography. (1)

1.2 Mixtures and compounds

(a) State three differences between a mixture and a compound. (7)
(b) Describe how you would separate iron filings and powdered sulphur from a mixture of the two. (4)
(c) (i) What would you *observe* if you heated a mixture of iron filings and sulphur in a test-tube? (4)
(ii) Describe one chemical test by means of which you could show that a new compound had been formed. (3)

1.3 Solubility and separation of gases

(a) (i) Name two gases that are very soluble in water and two gases that are almost insoluble in water. (4)
(ii) Suggest an explanation for the high solubility of the first two gases named. (1)
(b) Draw a labelled diagram to illustrate the method you would use to remove water vapour from a stream of moist oxygen formed in a chemical reaction. (The methods of preparation and collection of oxygen are not required.) (2)
(c) Describe, with the aid of a diagram, how you would attempt to determine the approximate percentage by volume of carbon dioxide in a mixture of carbon dioxide and nitrogen. (7)

1.4 Separation of mixtures

Describe briefly how you would extract a pure sample of the substance printed in bold type from each of the following mixtures:
(a) ammonium chloride and sodium chloride, (3)
(b) calcium carbonate and sodium carbonate, (4)
(c) copper powder and zinc powder, (4)
(d) nitrogen and ammonia. (3)

1.5 Separation of a mixture

(a) Describe in detail how you would produce pure, dry samples of copper(II) oxide, iodine and sand from a mixture of all three. (12)
(b) A student was trying to make pure, dry copper(II) sulphate crystals from the solution obtained by reacting copper(II) oxide with dilute sulphuric acid. Instead of blue crystals, his final product was a hard grey-white solid.
(i) What mistake had he made? (2)
(ii) How would you avoid the error? (4)

2 Particles, diffusion and radioactivity

2.1 Diffusion of gases 1

An experiment was set up as shown in the diagram below.

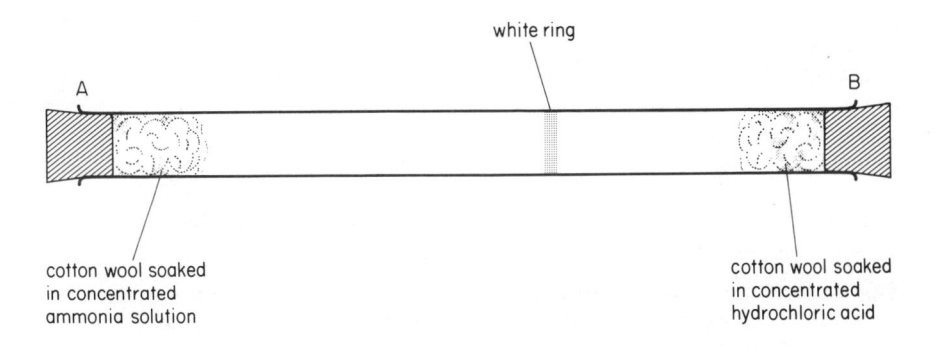

white ring

A
B

cotton wool soaked in concentrated ammonia solution

cotton wool soaked in concentrated hydrochloric acid

After several minutes a white ring formed in the position shown.
- **(a) (i)** What is the name of the compound that forms the white ring? (1)
  - **(ii)** Write the chemical equation for the reaction by which the compound is formed. (1)
- **(b)** Explain why the ring was formed nearer to the hydrochloric acid than to the ammonia solution. (2)
- **(c)** Explain why the tube must be
  - **(i)** level. (1)
  - **(ii)** corked. (2)
- **(d)** After the ring had formed, the corks and pieces of cotton wool were removed and the tube heated by passing a Bunsen burner slowly from right to left along the tube. As the flame passed under the white ring it disappeared and reappeared further to the right. What reason can you give for the disappearance and reappearance of the ring in this part of the experiment? (1)
- **(e)** Where would you expect the ring to form if, in a similar experiment:
  - **(i)** the cotton wool at end A had been soaked in hydrobromic acid (HBr) instead of hydrochloric acid? (1)
  - **(ii)** the tube had been heated very gently at end B immediately after the apparatus shown in the diagram was set up? (1)
  In each case explain your answer. (2)

2.2 Diffusion of gases 2

The time taken for 100 cm³ of each of the following gases to diffuse was measured. The results were as follows:

Gas	Formula	Relative molecular mass (M_r)	Time taken for 100 cm³ to diffuse
helium	T	4	24 seconds
methane	CH_4	U	48 seconds
sulphur dioxide	SO_2	64	96 seconds
hydrogen bromide	V	W	108 seconds
an alkane	X	100	120 seconds

(a) In the table there are five spaces labelled T, U, V, W and X.
  (i) What are the formulae for T, V and X? (3)
  (ii) What are the M_r values for U and W? (2)
(b) Draw a graph of the time taken for 100 cm³ of each gas to diffuse against M_r of the gas. (5)
(c) Three other gases, A, B and C, each took 64 seconds to diffuse. What is the M_r value for these gases? Identify each of the gases from the information given.
   Gas A burns with a bright blue flame to form as the only product a gas that turns lime water milky.
   Gas B shows no positive tests.
   Gas C is a hydrocarbon that decolourises bromine water. (4)
(d) (i) What is the M_r of the halogen that takes 101 seconds to diffuse? (1)
   (ii) Identify the halogen. (1)

2.3 Radioactive decay series

Some radioactive elements decay by the emission of alpha particles, and some by the emission of beta particles, into other elements which are themselves radioactive. This process continues until a non-radioactive element is produced. Such a series is shown below.

$^{238}_{92}U \xrightarrow{\alpha} {}^{234}_{90}Th \longrightarrow {}^{234}_{91}Pa \xrightarrow{\beta} {}^{234}_{92}U \longrightarrow Y \longrightarrow {}^{226}_{88}Ra \longrightarrow {}^{222}_{86}Rn \xrightarrow{\beta} Z \longrightarrow {}^{218}_{85}At$

(a) Identify, giving a reason in each case, the radioactive particles emitted by:
  (i) $^{234}_{90}Th$ (2)   (ii) $^{222}_{86}Rn$ (2)
(b) What name is given to atoms with different mass numbers but with the same atomic number?   (1)
(c) Identify the element Y, giving its mass number, atomic number and symbol.   (3)
(d) The half-life of $^{234}_{90}Th$ is 24 days.
  (i) What is meant by the term 'half-life'?   (1)
  (ii) What fraction of the original number of radioactive atoms will be left after 96 days?   (1)
(e) $^{222}_{86}Rn$ is a noble gas. In which Group of the Periodic Table is Z? Explain your answer.   (2)
(f) How many alpha particles and beta particles will need to be emitted to obtain the stable element $^{206}_{82}Pb$ from $^{218}_{85}At$?   (4)

2.4 Radioactive elements

(a) What is meant by the following statement?
'There are three common isotopes of uranium: their mass numbers are 234, 235 and 238.'   (4)
(b) When 238-U decays, it emits an alpha particle. The daughter atom so formed is thorium (Th) and this emits a beta particle.
  (i) What are beta particles?   (1)
  (ii) Complete the following nuclear equations by putting numbers in place of each star:

  $^{238}_{92}U \longrightarrow ^{*}_{*}He + ^{*}_{*}Th$

  $^{*}_{*}Th \longrightarrow ^{*}_{*}e^{-} + ^{*}_{*}Pa$ (8)

(c) When uranium is used as a fuel in a nuclear power plant the uranium is used up, but it is not 'burned' in the same way as coal or oil in a conventional plant. How does the uranium release energy?   (2)

2.5 Nuclear reactions

The following equations represent nuclear reactions:

$^{9}_{4}Be + ^{4}_{2}He \longrightarrow ^{12}_{6}C + ^{1}_{0}n$ (I)

$^{14}_{7}N + ^{4}_{2}He \longrightarrow ^{17}_{8}O$ (II)

(a) By what names are the particles $^{4}_{2}He$ and $^{1}_{0}n$ usually known?   (2)
(b) (i) Explain why equation (II) is, unlike equation (I), incomplete.   (4)
  (ii) Rewrite equation (II) putting in the missing particle so that the equation is balanced.   (3)
(c) Explain why an element may have a relative atomic mass (A_r) which is not a whole number, although each atom must contain a whole number of protons and a whole number of neutrons.   (4)

3 Atomic structure and chemical bonds

3.1 Atomic structure and bonding

The diagram below shows the electronic arrangement of an atom of an element X.

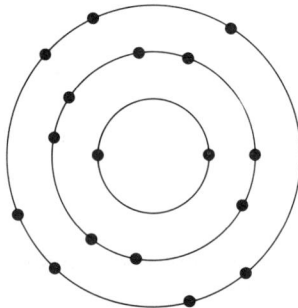

(a)   **(i)** Is X a metal or a non-metal?   (1)
   (ii) How do you know this?   (2)
   (iii) Draw a diagram to show the arrangement of electrons in an atom of a noble gas in the same Period of the Periodic Table as element X.   (2)
(b) X is known to form a covalent bond with hydrogen.
   (i) Write the molecular formula of this compound.   (2)
   (ii) Draw a diagram of the electronic arrangement in a molecule of the compound.   (2)
(c) Study the following list of properties:
high melting point, low melting point, soluble in water, soluble in ethanol, conducts electricity when molten, does not conduct electricity when liquid.
   (i) Rewrite these properties in two lists, one headed 'Ionic compounds' (e.g. sodium chloride), and the other 'Simple covalent compounds' (e.g. methane). Put each property into the appropriate column.   (3)
   (ii) Place each of the following substances into the correct column: copper(II) sulphate, ethene, sugar, rubidium fluoride.
(Rubidium is an element with atomic number 37.)   (4)

3.2 Atomic structure 1

(a) The table below gives information about the lithium atom, the oxygen ion and the magnesium atom.

	Mass number	Atomic number	Number of protons	Number of electrons	Number of neutrons
lithium atom	7	3	3	A	B
oxygen ion O²⁻	16	C	8	D	E
magnesium atom	F	G	H	12	12

What are the missing values A, B, C, D, E, F, G and H?   (4)
- **(b)** **(i)** What do you understand by the word 'isotope'?   (2)
  - **(ii)** Bromine exists as two isotopes of mass numbers 79 and 81. The two isotopes are equally abundant. What is the M_r of bromine?   (2)
- **(c)** Lithium reacts with fluorine to form the ionic solid lithium fluoride.
  - **(i)** Draw diagrams to show the full electronic configuration of the lithium atom, the fluorine atom, the lithium ion and the fluoride ion.   (4)
  - **(ii)** Give three typical properties of ionic compounds.   (3)
  - **(iii)** Name a compound with similar properties to lithium fluoride.   (1)

3.3 Atomic structure 2

- **(a)** An atom of an element X has an atomic number of 11 and a mass number of 23. Give the number of protons, neutrons and electrons in an atom of X.   (3)
- **(b)** Using the two isotopes of chlorine, $^{35}_{17}Cl$ and $^{37}_{17}Cl$, explain
  - **(i)** why the two isotopes have the same chemical properties   (2)
  - **(ii)** why the relative atomic mass is not a whole number.   (2)
- **(c)** Draw a diagram to show the arrangement of electrons in the compound formed between X and chlorine.   (4)
- **(d)** The chloride of a non-metal Z reacts with water to give hydrogen chloride and the compound $Z(OH)_3$.
  - **(i)** Describe how you would prove that the gas given off was hydrogen chloride.   (2)
  - **(ii)** Suggest the formula of the chloride of Z. Explain your answer.   (2)
  - **(iii)** In which group of the Periodic Table would you expect to find element Z? Explain your answer.   (2)

3.4 Atomic number and bonding

W, X and Y are three elements with atomic masses n, $n + 1$ and $n + 2$. X is a chemically inert gas. W and Y react together to form a solid compound T.
- (a) (i) What type of bonding will be present in the compound formed between W and Y? (1)
  - (ii) Give the formula of the compound formed between W and Y. (1)
  - (iii) Would you expect this compound to conduct electricity? Explain your answer. (3)
  - (iv) Give two properties of this compound, other than electrical properties. (2)
- (b) (i) If magnesium forms a compound with W, what particles would be present in the compound? (2)
  - (ii) Give the formula of this compound. (1)
- (c) What is the atomicity of
  - (i) element W (1)
  - (ii) element X? (1)
- (d) Name an element in the same Group of the Periodic Table as X. (1)
- (e) 13 g of Y reacts with excess cold water to give 4 dm³ of hydrogen at room temperature and pressure. Identify Y and hence identify W and X. (6)

3.5 Chemical bonding

The diagram below illustrates, in terms of the arrangement of electrons within atoms and ions, the combination of lithium and fluorine to form lithium fluoride.

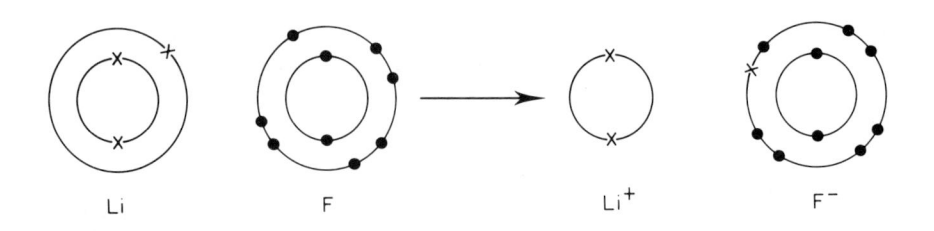

- (a) (i) In a similar way draw an electron diagram for the formation of sodium chloride. (4)
  - (ii) What type of bond is present in sodium chloride? (1)
- (b) What would you expect to take place if sodium chloride is added to
  - (i) water (1)
  - (ii) ethanol? (1)

(c) An experiment was performed on a compound X, containing only carbon and chlorine, to find the molecular formula of X. It was found that 0.0042 mole of X contained 0.10 g of carbon and 0.90 g of chlorine.
  (i) Calculate the molecular formula of X. (4)
  (ii) Draw a diagram to show the arrangement of electrons in X. (Show only the outermost electrons.) (3)
  (iii) What type of bonding is present in X? (1)
  (iv) What would you expect to take place if X is added to water and to ethanol? (2)

3.6 Structure and bonding

Chloride	Boiling point/°C	Chloride	Boiling point/°C
HCl	−85	neon	no chloride
helium	no chloride	NaCl	1460
LiCl	1380	$MgCl_2$	1410
$BeCl_2$	550	$AlCl_3$	420
boron chloride	15	$SiCl_4$	60
CCl_4	80	PCl_3	80
nitrogen chloride	70	S_2Cl_2	−80
OCl_2	2	Cl_2	−34
FCl	−90	argon	no chloride

(a) Which two chlorides have identical boiling points? (1)
(b) Suggest the formulae of:
  (i) boron chloride (1)
  (ii) nitrogen chloride. (1)
(c) Explain why helium, neon and argon do not form chlorides. (2)
(d) Select from the table three chlorides which will be gaseous at room temperature and pressure. Give one reason for choosing these chlorides. (2)
(e) Which chlorides in the table will be solids at 1400°C? (1)
(f) Using FCl and $MgCl_2$ as your examples, explain what is meant by
  (i) covalent bonding (1)
  (ii) ionic bonding. (1)
(g) Predict the boiling points of
  (i) potassium chloride (1)
  (ii) arsenic chloride. (1)

(*contd.*)

(h) S_2Cl_2 does not fit the pattern. S_2Cl_2 reacts with water to give a pale yellow solid A, a colourless solution and a gas B that fumes in moist air. When aqueous barium chloride is added to the colourless solution, a white precipitate is formed which is soluble in excess dilute hydrochloric acid.
  (i) What is the expected formula for the chloride of sulphur? (1)
  (ii) Identify A and B and hence construct the equation for the reaction between S_2Cl_2 and water. (4)

4 Moles and equations

4.1 Mole calculation

A compound of lead and chlorine contains 59.3% by mass of lead.
(a) (i) Calculate the mass of chlorine combined with one mole of lead. (3)
  (ii) Work out the simplest formula of the compound. (2)
  (iii) What is the M_r of the compound? (1)
(b) One-twentieth (1/20) of a mole of the compound was decomposed into its elements at room temperature and pressure.
  (i) What mass of lead was produced? (2)
  (ii) What volume of chlorine was produced? (3)
(c) The chlorine was then completely converted into tetrachloromethane, CCl_4.
  (i) How many moles of tetrachloromethane were produced? (1)
  (ii) What was the mass of the product? (2)

4.2 The mole/oxides of nitrogen

(a) 1.25 g of a gaseous oxide of nitrogen has a volume of 1.00 dm³ at room temperature and pressure. When the oxide is passed over heated copper, it is completely reduced.
  (i) Deduce the formula of the oxide of nitrogen. (2)
  (ii) Construct the equation for the reaction of copper with this oxide of nitrogen. (1)
  (iii) What volume of nitrogen is obtained when 1.00 dm³ of the oxide of nitrogen is passed over excess hot copper? Name and state any law that you have used. (5)
(b) Nitrogen dioxide, NO_2, is formed when lead(II) nitrate is heated. Nitrogen dioxide reacts with water to form a mixture of nitric acid and nitrous acid (HNO_2).

(i) Name the products formed when excess nitrogen dioxide is bubbled into aqueous sodium hydroxide. Write the equation for this reaction. (3)

(ii) What volume of 0.1 mol/dm^3 sodium hydroxide solution would exactly react with 1.2 dm^3 of nitrogen dioxide measured at room temperature and pressure? (3)

4.3 The mole/energy (enthalpy) change

Various masses of a metal X were added to separate 50 cm^3 portions of 0.2 mol/dm^3 copper(II) sulphate solution contained in a polystyrene cup. The maximum rise in temperature was noted in each case. The results were:

Mass of X used/g	Maximum rise in temperature/°C
0.1	2.5
0.2	5.0
0.3	7.5
0.4	9.8
0.5	12.3
0.6	13.5
0.7	13.5

(a) Plot a graph of these results, plotting maximum rise in temperature on the y-axis. (5)
(b) Explain the shape of the graph you obtained in (a). (3)
(c) From the graph find out the mass of X that exactly reacts with 50 cm^3 of 0.2 mol/dm^3 copper(II) sulphate solution. (1)
(d) The equation for the reaction of X with copper(II) sulphate solution is:

$$X(s) + CuSO_4(aq) \longrightarrow Cu(s) + XSO_4(aq)$$

(i) Calculate the A_r of X. (2)
(ii) Use the Periodic Table to identify element X. (1)
(e) Describe what you would observe if excess X were added to copper(II) sulphate solution. (2)
(f) Why was the experiment carried out in a polystyrene cup? (1)
(g) When excess zinc was added to 50 cm^3 of 0.2 mol/dm^3 copper(II) sulphate solution, the rise in temperature was 10°C. What can you deduce from this result about metal X? (1)

4.4 The mole/potassium manganate(VII)

Solutions of potassium manganate(VII) ($KMnO_4$) and hydrogen peroxide (H_2O_2) react rapidly in the presence of dilute sulphuric acid according to the equation:

$$2KMnO_4(aq) + 3H_2SO_4(aq) + 5H_2O_2(aq) \longrightarrow$$
$$K_2SO_4(aq) + 2MnSO_4(aq) + 8H_2O(l) + 5O_2(g)$$

The salt $MnSO_4$ is colourless in dilute solution.
A is a solution containing 0.1 mole of $KMnO_4$ per dm^3.
B is a solution containing 0.1 mole of H_2O_2 per dm^3.

(a) What is the minimum volume of B required to react completely with 100 cm^3 of A?   (3)
(b) Write the formulae of **all** the ions present in a solution containing the salts K_2SO_4 and $MnSO_4$.   (5)
(c) What would be the final colour of the mixture if you mixed together the following quantities of solution in the presence of dilute sulphuric acid? In each case give a reason for your answer.
   (i) 100 cm^3 of A and 300 cm^3 of B.   (2)
   (ii) 100 cm^3 of A and 30 cm^3 of B.   (2)
(d) What is the maximum volume of oxygen, at room temperature and pressure, that could be obtained from 100 cm^3 of B by treating it with potassium manganate(VII) and dilute sulphuric acid?   (3)
(e) When manganese(IV) oxide is added to hydrogen peroxide, oxygen is given off.
   (i) What is the purpose of manganese(IV) oxide in this reaction?   (1)
   (ii) Write the equation for this reaction.   (1)
   (iii) What is the maximum volume of oxygen, measured at room temperature and pressure, that can be obtained from 100 cm^3 of B by this reaction?   (2)

4.5 Mole calculation/precipitation

Several 10 cm³ portions of 0.1 mol/dm³ of a metal sulphate were put into test-tubes and each was mixed with a different volume of 0.2 mol/dm³ barium chloride solution. The resulting precipitate of barium sulphate (BaSO$_4$) was filtered, washed, dried and weighed. The results are set out below:

Experiment	A	B	C	D	E	F
Volume of metal sulphate/cm³	10	10	10	10	10	10
Volume of barium chloride/cm³	2.5	5.0	10.0	15.0	20.0	25.0
Mass of barium sulphate/g	0.1165	0.2330	0.4660	0.6990	0.6990	0.6990

(a) Write the ionic equation for the reaction between barium chloride and the metal sulphate. (1)
(b) Calculate the M_r of barium sulphate. (1)
(c) Why did the mass of the precipitate remain constant in experiments D, E and F? (1)
(d) (i) Calculate the number of moles of metal sulphate in 10 cm³ of the solution. (1)
  (ii) Calculate the number of moles of barium chloride which reacted exactly with 10 cm³ of the metal sulphate. (1)
  (iii) Using your answer to **(b)**, calculate the number of moles of barium sulphate formed. (1)
(e) (i) If X$_x$(SO$_4$)$_y$ is the formula of the metal sulphate, what is the value of y? (2)
  (ii) What is the value of x? (1)
(f) (i) The M_r of X$_x$(SO$_4$)$_y$ is 400. What is the A_r of X? (2)
  (ii) Identify X and suggest two physical properties of the metal sulphate X$_x$(SO$_4$)$_y$. (3)

5 Electrolysis and electrical cells

5.1 Conductors/non-conductors

The apparatus shown in the diagram below was set up to test the chemical changes that take place during the electrolysis of dilute sulphuric acid using different electrodes.

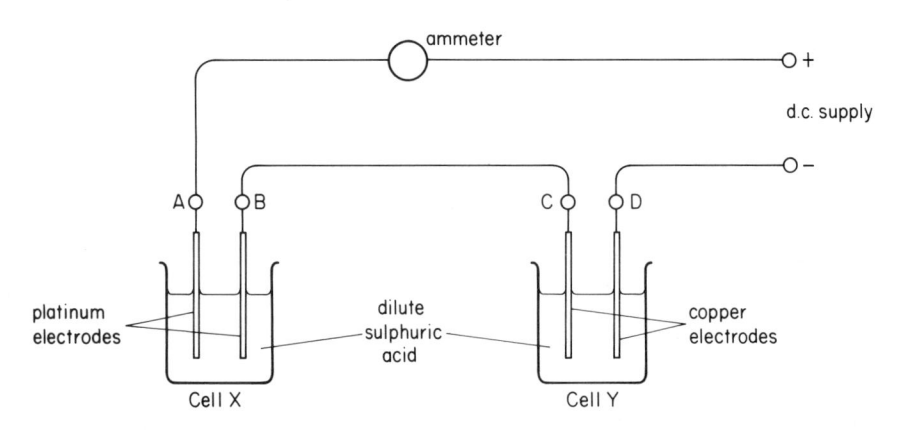

(a) Give the letters of the positive electrodes. (2)
(b) What gases are formed in cell X? (2)
(c) In cell Y no gas is given off at electrode C; instead it decreases in mass. Explain this observation. (3)
(d) Name the gas formed at electrode D in cell Y. Give the electrode equation for the formation of this gas. (2)
(e) Describe what you would expect to be formed in (i) cell X and (ii) cell Y if the dilute sulphuric acid were replaced in both cells by each pair of the following substances. The electrodes are unchanged.

Experiment	Cell X	Cell Y
1	mercury	a concentrated solution of sodium chloride in water
2	ethanol	dilute hydrochloric acid
3	solid sodium chloride	molten sodium chloride
4	molten sodium chloride	a concentrated solution of sodium chloride in water

(11)

5.2 Ions

The following table gives information about elements, the relative atomic masses of the elements, the electrode at which the element is discharged, the mass of the element deposited by 1 Faraday (96 500 coulombs) and the charges on the ions.

Element	Relative atomic mass	Electrode	Mass in grams per Faraday	Formula of ion
aluminium	27.0	A	B	Al^{3+}
chlorine	35.5	C	D	Cl^-
E	1.0	cathode	1.0	F
oxygen	G	anode	8.0	H
samarium	I	J	50.0	Sm^{3+}
K	45.0	cathode	15.0	L
tin	M	N	60.0	Sn^{2+}

Examine the above table and identify A, B, C, D, E, F, G, H, I, J, K, L, M and N. (14)

5.3 Electrolysis of aqueous solutions

The diagram below shows the electrolysis of dilute sulphuric acid.

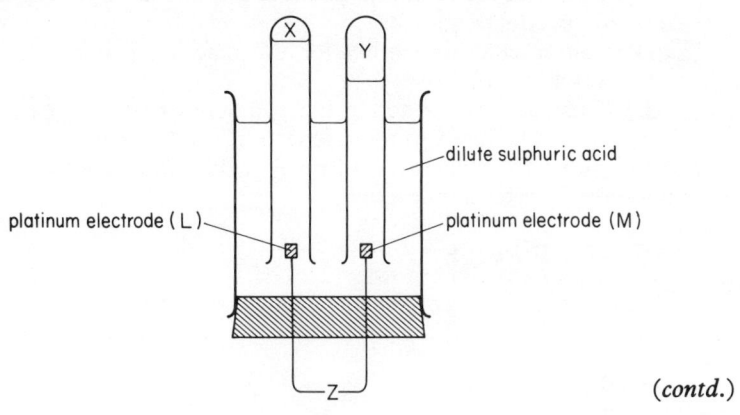

(contd.)

(a) (i) If the experiment is to work, what must be placed in the circuit at Z? (1)
   (ii) *Name* the gases X and Y. (2)
   (iii) Which is the positive electrode, L or M? Explain how you arrived at this answer. (2)
(b) (i) Write the formulae of all the ions present in the electrolyte. (3)
   (ii) Write ionic equations to show what happens at each electrode. (2)
(c) If the apparatus is emptied and then filled with a solution of copper(II) sulphate, what would you see happening at each electrode? (4)
(d) After the apparatus had been used for some time with copper(II) sulphate solution as the electrolyte, the electrodes were removed, rinsed with water, and immersed separately into concentrated nitric acid. What would you observe in each case? (3)

5.4 Electricity and fluorine

(a) Which of the following substances conduct electricity:
   (i) when solid (1)
   (ii) when molten, but not when solid (1)
   (iii) when dissolved in water (1)

| aluminium oxide | ammonia | hydrogen chloride |
| iron | silver chloride | sodium |

(b) Briefly describe how you would silver-plate a copper spoon. State the precautions you would take to ensure good electroplating. (5)
(c) Fluorine is extracted by electrolysis of potassium fluoride. What volume of fluorine, at room temperature and pressure, is formed when a current of 1.0 ampere is passed for 1 hour through molten potassium fluoride? (4)
(d) Fluorine reacts with water to produce oxygen and hydrofluoric acid as the only products.
   (i) Construct the equation for the reaction between fluorine and water. (1)
   (ii) What does this reaction tell you about the strength of the H–F bond compared to the strength of the O–H bond? (1)
(e) Write down the structural formula of the monomer from which the polymer poly(tetrafluoroethane)

$$-\underset{F}{\overset{F}{C}}-\underset{F}{\overset{F}{C}}-\underset{F}{\overset{F}{C}}-\underset{F}{\overset{F}{C}}-\underset{F}{\overset{F}{C}}-\underset{F}{\overset{F}{C}}-$$

is made. (2)

5.5 Electrolysis of copper(II) sulphate

100 cm³ of a solution containing 0.50 mol/dm³ of copper(II) sulphate was poured into a U-tube. Carbon electrodes A and B were placed in the solution and the apparatus set up as shown in the diagram below.

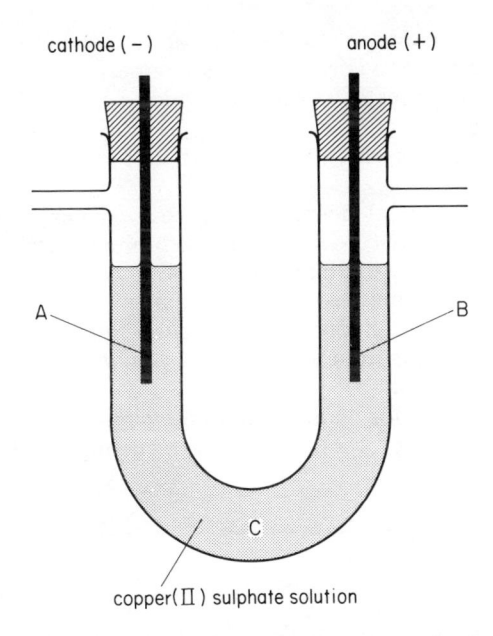

(a) What would you observe at electrode A after a few minutes? (1)
(b) Bubbles of a colourless gas were produced at electrode B. This gas turned lime water milky.
  (i) Name the gas that turned lime water milky. (1)
  (ii) Explain how this particular gas was formed at electrode B. (3)
  (iii) If this gas were passed through water, what would be the effect on the pH of the water? (1)
(c) After 30 minutes the current was switched off. Describe the appearance of the solution at C. Give a reason for your answer. (2)
(d) Electrode B was replaced by a piece of platinum. What difference will this have on the products of the electrolysis? (2)
(e) In a similar experiment, both electrodes were made of copper and weighed at the start of the experiment. A steady current of 0.1 ampere was allowed to pass for 30 minutes. When the electrodes were disconnected, dried and weighed, the mass of one of them was found to have increased by X g.
  (i) How many coulombs passed through the circuit? (1)
  (ii) What is the value of X? (2)
  (iii) What change in mass would you expect in the other copper electrode? (1)
  (iv) What would be the concentration of the copper(II) sulphate solution at the end of the experiment? (1)

5.6 Transition metals/electroplating

Chromium is a typical transition metal. It is manufactured by heating chromium(III) oxide with aluminium. In the activity series it is placed between zinc and iron. It forms two soluble sulphates, $CrSO_4$ and $Cr_2(SO_4)_3$.

Chromium(III) oxide is used as a catalyst for the manufacture of methanol from hydrogen and carbon monoxide.

(a) Give two properties of chromium which suggest that it is a typical transition metal.   (2)
(b) Describe, with reasons, the reactions that you would expect to take place if
   (i) chromium metal is added to aqueous copper(II) sulphate   (2)
   (ii) chlorine is passed over heated chromium metal.   (2)
(c) Write equations for the manufacture of
   (i) chromium from chromium(III) oxide   (1)
   (ii) methanol from hydrogen and carbon monoxide.   (1)
(d) Draw a labelled diagram to show how you would electroplate a metal object with chromium.   (5)
(e) Chromium is used for plating iron to prevent it from rusting. Give two other methods of preventing iron from rusting.   (2)

5.7 Redox/cells

The apparatus shown in the diagram below was set up.

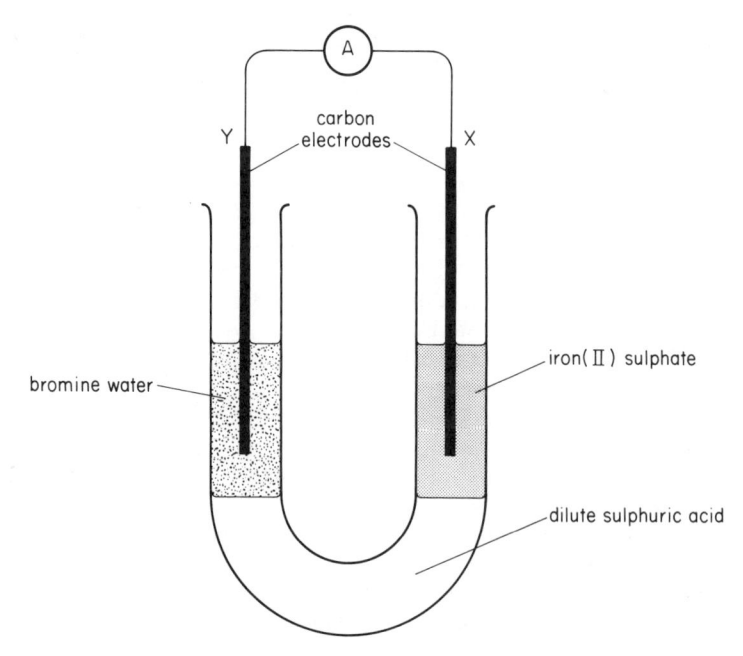

Electrons flowed through the external circuit from electrode X to electrode Y. The current was detected by the ammeter A. After some time, the bromine water was decolourised, and the iron(II) sulphate had turned from a green solution to a brown solution.
- **(a)** **(i)** Why could the electrode Y not be made of iron? (2)
  - **(ii)** Why could the electrode X not be made of zinc? (2)
- **(b)** What are the final products in
  - **(i)** the left-hand side of the U-tube (1)
  - **(ii)** the right-hand side of the U-tube? (1)
  - **(iii)** How would you confirm your answers to **(b) (i)** and **(ii)**? (4)
  - **(iv)** Write down the ionic equations for the electrode reactions. (2)
- **(c)** Using your answer to **(b)(iv)**, or otherwise, write down the equation for the reaction between bromine water and iron(II) ions. (2)
- **(d)** State, with a reason, how the flow of current would change if sugar solution were used instead of dilute sulphuric acid. (2)

6 Energy changes, speeds of reaction and reversible reactions

6.1 Chemical reactions

Give one example of each type of chemical reaction in the following list. In each case state the products and the reactants, and either name the process or give a use to which it can be put.
- **(a)** Light energy is given out. (3)
- **(b)** Light energy is absorbed. (3)
- **(c)** Electrical energy is given out. (3)
- **(d)** Heat energy is given out. (3)
- **(e)** Heat energy is absorbed. (3)

6.2 Speed of reaction 1

When sodium thiosulphate solution reacts with dilute hydrochloric acid, the solution slowly becomes cloudy because sulphur is precipitated.

A student has a solution of sodium thiosulphate, a quantity of dilute hydrochloric acid, and a supply of distilled water. He mixes 40 cm^3 of sodium thiosulphate with 40 cm^3 of the acid in a beaker on a sheet of paper on which is drawn an X, and notes the time interval between mixing the solutions and the mixture becoming so cloudy that he can no longer see the X. *(contd.)*

He then repeats the experiment several times, keeping the total volume the same but diluting the sodium thiosulphate solution with distilled water before mixing it with acid.

His measurements are shown in the table. Use these data to answer the questions below.

Volume of sodium thiosulphate solution /cm^3	Volume of distilled water /cm^3	Volume of hydrochloric acid /cm^3	Time (t) for X to vanish /second	$1/t$ /second^{-1}
40	0	40	8	0.125
30	10	40	11	0.091
20	20	40	14	0.071
15	25	40	21	0.048
10	30	40	32	0.031

(a) **(i)** Is the speed of the reaction the same for each experiment? (1)
   (ii) Draw a graph of 1/(vanishing time) i.e. $1/t$, on the y-axis, against the volume of thiosulphate used. (5)
   (iii) One of the times recorded is probably wrong. Which one is it? How do you know? (2)
(b) What volumes of thiosulphate and water would have to be used to produce a vanishing time of 26 seconds? (2)
(c) Give two ways of speeding up this reaction. (2)
(d) **(i)** Name the gas given off when sodium thiosulphate reacts with hydrochloric acid. (1)
   (ii) Suggest another method of studying the speed of this reaction. (1)

6.3 Speed of reaction 2

The graph below shows the total volume of gas produced in a certain reaction, plotted against time.

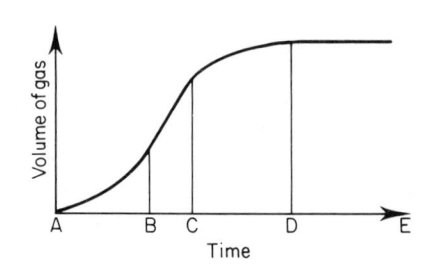

(a) (i) During which interval of time was the speed of reaction increasing? (1)
(ii) During which interval of time was the speed of reaction constant? (1)
(iii) When did the reaction stop? (1)
(b) (i) Suggest what might have caused the reaction speed to increase. (2)
(ii) Why did the reaction speed eventually decrease? (2)
(iii) Why did the reaction stop? (2)
(c) (i) Draw a diagram of the apparatus you would use to measure the volume of hydrogen produced in the reaction between zinc and dilute hydrochloric acid. (5)
(ii) State one precaution you would take to make sure that the experiment was done safely. (1)

6.4 Speed of reaction 3

Two portions of calcium carbonate, each weighing 5.0 g, both consisted of the same sized particles. One portion was added to 1.0 dm^3 of dilute hydrochloric acid of concentration 1.0 mol/dm^3. The other portion was added to 100 cm^3 of identical acid. The graphs below show the total volume of gas given off from each sample, plotted against time.

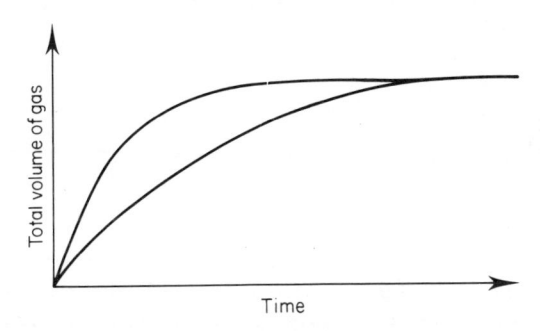

(a) Calculate the number of moles of calcium carbonate in 5.0 g. (2)
(b) Why do both curves come to the same final volume? (5)
(c) What would be the final volume of gas, measured at room temperature and pressure? (4)
(d) What do the different slopes of the two graphs near the origin tell you about the rates of the two reactions? (1)
(e) Suggest the reason for this difference. (1)
(f) Why do the slopes become less and less steep as time goes on? (4)

6.5 Speed of reaction 4

Experiments 1, 2 and 3 were set up to investigate the effect of concentration of an acid on the rate of its reaction with magnesium. The total volume of
(contd.)

hydrogen produced was measured at regular time intervals. The experiments were carried out with three different concentrations of acid, as shown in the table.

Experiment	Mass of magnesium/g	Volume of acid/cm^3	Concentration of acid/mol dm^{-3}
1	0.24	40	0.5
2	0.24	40	1.0
3	0.24	40	2.0

In each experiment freshly cleaned magnesium ribbon was used. The results of the experiments are shown below.

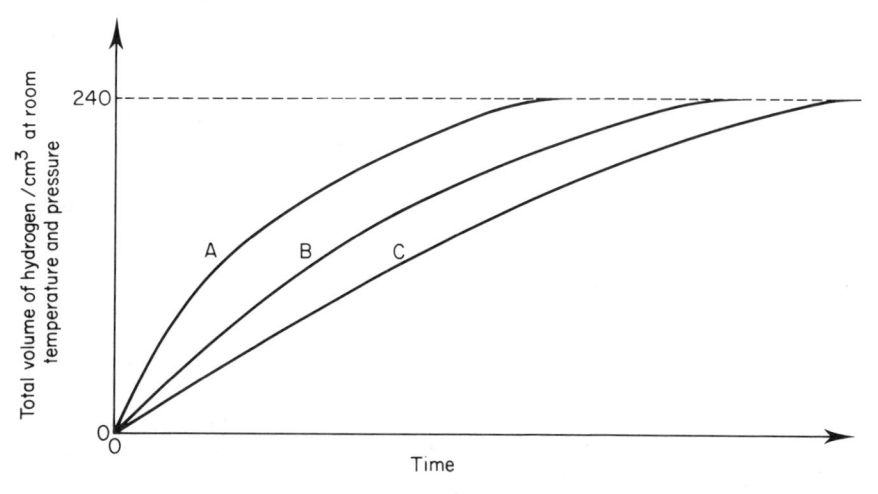

(a) (i) To which experiments do curves A and B refer? (2)
 (ii) Which curve refers to the reaction with the greatest initial speed of reaction? Give a reason for your answer. (2)
(b) Suggest why freshly cleaned magnesium ribbon was used in each case. (2)
(c) From the information given above, and given that **all** the magnesium and **all** the acid reacted in experiment 1, deduce the ionic equation for the reaction by answering the following questions.
 (i) How many moles of magnesium were used? (1)
 (ii) How many moles of acid were used? (1)
 (iii) How many moles of hydrogen were given off? (1)
 (iv) Using the values obtained in (i), (ii) and (iii), deduce the ionic equation for the reaction between magnesium and the acid. (2)
(d) Copy the graphs in the diagram above and the curve (D) that might be obtained by using 40 cm^3 of a 0.25 mol/dm^3 solution of the acid, keeping the other conditions the same. Explain the shape of your graph. (3)

6.6 Catalysts

(a) Give a definition of a catalyst. (2)
(b) State three reactions in which an inorganic catalyst is used. (These may be laboratory or industrial processes.) For each one, say
  (i) what the reactants are (3)
  (ii) what the products are (3)
  (iii) what the catalyst is. (3)
(c) What do we call the catalysts involved in life processes? What are the essential conditions before such a catalyst will work? (3)

6.7 Catalyst: chemical equilibrium

(a) Give a definition of a catalyst. (2)
(b) The **Haber** process converts nitrogen and hydrogen into ammonia:

$$N_2(g) + 3H_2(g) \rightleftharpoons 2NH_3(g) \qquad \Delta H = -95 \text{ kJ/mol}$$

  (i) What catalyst is used in this process? (1)
  (ii) Does the reaction give out or absorb heat energy? (1)
(c) If the temperature of the reaction is increased, what will be the effect on
  (i) the speed of the reaction (1)
  (ii) the position of the equilibrium? (1)
(d) What will be the effect on the equilibrium position of adding a catalyst? Explain your answer. (3)
(e) To get the best yield, would you do the reaction at a high or a low pressure? Explain your reasoning. (2)

6.8 Reversible reactions

The reaction between iron(II) ions and silver ions is reversible:

$$Fe^{2+}(aq) + Ag^+(aq) \rightleftharpoons Fe^{3+}(aq) + Ag(s)$$

(a) Describe one chemical test in each case to show the presence of:
  (i) $Fe^{2+}(aq)$ (2)
  (ii) $Fe^{3+}(aq)$ (2)
  (iii) $Ag^+(aq)$ (2)
(b) Describe what you would observe if iron(II) sulphate solution were shaken with silver nitrate solution. (2)
(c) Pure nickel is obtained by heating tetracarbonylnickel, $Ni(CO)_4$, to 200°C. The reaction is reversible.

$$Ni(CO)_4(s) \rightleftharpoons Ni(s) + 4CO(g)$$

Calculate the mass of nickel that can be obtained from 19 g of $Ni(CO)_4$, assuming that 90% of the $Ni(CO)_4$ is decomposed at 200°C. (4)
(d) Name two compounds that are manufactured by processes which involve reversible reactions. (2)

7 Salt preparations, strong and weak acids

7.1 Salt preparation: copper(II) sulphate

The following account correctly describes a method for the preparation of copper(II) sulphate crystals, ($CuSO_4 \cdot 5H_2O$).

Pour about 50 cm³ of bench sulphuric acid into a small beaker. Warm the acid but do not boil it. Add copper(II) oxide, a spatula load at a time, while stirring, until the copper(II) oxide is in excess. Filter off the copper(II) oxide and heat the filtrate to crystallising point. Leave the filtrate to cool. When the crystals have formed, filter off the crystals and dry them with filter paper.

(a)   (i) Why does it not matter if the amount of bench sulphuric acid used was more or less than 50 cm³? (1)
  (ii) Why is the mixture stirred with a glass rod and not with a nickel spatula? (1)
  (iii) Why are the crystals dried between filter papers and not heated carefully to dryness? (1)
  (iv) Why is the sulphuric acid heated? (1)

(b)   (i) Write the equation for the reaction between copper(II) oxide and sulphuric acid. (1)
  (ii) Calculate the maximum mass of copper(II) sulphate crystals that could be obtained from 50 cm³ of 1.0 mol/dm³ sulphuric acid. (Relative formula mass of $CuSO_4 \cdot 5H_2O$ is 250.) (3)
  (iii) When the experiment in (b) (ii) was carried out, the volume of filtrate left was 25 cm³. The filtrate was cooled to room temperature and the mass of copper(II) sulphate crystals obtained was 7.00 g. Calculate the solubility of copper(II) sulphate in water at room temperature. Express your answer as grams of $CuSO_4$ per 100 cm³ of water. (2)

(c) Name two salts manufactured on a large scale, and give one use for each salt. (4)

7.2 Salt preparation: calcium sulphate

A class of students is told that calcium sulphate and calcium hydroxide are both sparingly soluble in water. They are asked to suggest, in outline, a suitable method for the preparation of calcium sulphate from calcium carbonate. Four of their suggested methods are listed below.

Method A:
calcium carbonate $\xrightarrow{\text{dilute sulphuric acid}}$ calcium sulphate

Method B:
calcium carbonate →(heat) calcium oxide →(dilute sulphuric acid) calcium sulphate

Method C:
calcium carbonate →(dilute nitric acid) calcium nitrate solution →(dilute sulphuric acid) calcium sulphate

Method D:
calcium carbonate →(heat) calcium oxide →(water) calcium hydroxide solution →(dilute sulphuric acid) calcium sulphate

(a) Choose the method, A, B, C or D, that you think is the most suitable to obtain a good yield of calcium sulphate from calcium carbonate. Explain why each of the other three methods is less suitable. (5)
(b) Write equations for the reactions taking place in the method you have chosen. (2)
(c) How would you show the presence of Ca^{2+} ions in water? (2)

7.3 Ions

(a) Copy out the table below and complete it by writing in the formula of the ions present in an aqueous solution of each of the salts. (Ignore the ionisation of the water itself.)

Salt	positive ion	negative ion
ammonium chloride		
calcium chloride		
calcium hydrogencarbonate		
copper(II) sulphate		
iron(II) chloride		
silver nitrate		
sodium carbonate		
zinc nitrate		

(8)
(*contd.*)

(b) Write ionic equations to illustrate the reactions between solutions of
  (i) ammonium chloride and silver nitrate   (1)
  (ii) iron(II) chloride and sodium carbonate.   (1)
(c) Describe how you would distinguish between zinc nitrate solution and aqueous solutions of the other salts in the above table.   (3)

7.4 Strong acid

(a) (i) What is meant by a 'strong' acid?
  (ii) How do you explain that when hydrogen chloride is dissolved in methylbenzene (an organic liquid), the solution does not change the colour of litmus paper?   (3)
(b) Phosphoric(V) acid reacts with excess sodium hydroxide solution according to the equation

$$H_3PO_4(aq) + 3NaOH(aq) \longrightarrow Na_3PO_4(aq) + 3H_2O(l)$$

Calculate
  (i) the mass of the acid needed to react completely with 2.0 g of sodium hydroxide   (4)
  (ii) the concentration of the acid in mol/dm^3, if the mass in your answer to (b) (i) is dissolved in 200 cm^3 of water   (3)
  (iii) the number of moles of water produced in the reaction from 2 g of sodium hydroxide.   (1)
(c) Write an ionic equation which corresponds to the molecular equation given in (b).   (1)

7.5 Strong and weak acids

Hydrochloric acid is a strong acid, but ethanoic acid (CH_3CO_2H) is a weak acid.
(a) How do two solutions of these acids of concentration 2.0 mol/dm^3 differ in their
  (i) speed of reaction with magnesium   (2)
  (ii) electrical conductivity?   (2)
(b) Draw a diagram of the apparatus you would use to measure the speed of the reaction between hydrochloric acid and magnesium.   (4)
(c) What would be the effect on pH paper of the aqueous sodium salt of hydrochloric acid?   (1)
(d) Powder toilet cleaner contains sodium hydrogensulphate ($NaHSO_4$). This substance acts as a strong acid when dissolved in water. Suggest what you would observe on mixing separate portions of an aqueous solution of sodium hydrogensulphate with
  (i) magnesium   (1)
  (ii) sodium carbonate   (1)
  (iii) barium chloride.   (1)
(e) Construct the ionic equation for reaction (d)(i).   (1)
(f) Explain why it is dangerous to mix powder toilet cleaners with liquid toilet cleaners (which contain sodium chlorate(I)).   (2)

7.6 Organic acid

Solution A contains 0.05 mol/dm³ of a weak organic acid $(CO_2H)_n$. Solution B is a 0.10 mol/dm³ solution of sodium hydroxide. 20.0 cm³ of solution A requires 20.0 cm³ of solution B for neutralisation.

(a) What is meant by a 'weak' acid? (1)
(b) (i) Give the name and formula of a strong acid. (2)
 (ii) Explain the difference between a 'strong' acid and a 'concentrated' acid. (2)
(c) Calculate the number of moles of sodium hydroxide that are required to neutralise 1 mole of the weak acid. Hence calculate the value of n. (4)
(d) Write the equation for the reaction between the weak acid and sodium hydroxide. (1)
(e) Concentrated sulphuric acid dehydrates the weak acid. Suggest the products formed when concentrated sulphuric acid is added to $(CO_2H)_n$. (2)
(f) How would you show that the weak acid $(CO_2H)_n$ is a reducing agent? (2)

8 The Periodic Table

8.1 Periodic Table

(a) Look at the outline of part of the Periodic Table below.

Which part of the table, A, B or C, represents
  (i) the transition metals (1)
  (ii) Group I metals (1)
  (iii) the noble gases? (1)
(b) Give the names of
  (i) two transition metals, (1)
  (ii) two Group I metals (1)
(c) Using the metals you have named in (b), show how Group I metals and the transition metals differ with respect to:
  (i) melting point (2)
  (ii) the number of different oxidation states (numbers) (2)
  (iii) the colour of their compounds dissolved in water (2)
  (iv) their uses as catalysts (if any) (2)
  (v) effect of exposure of the metal to the atmosphere. (2)
(d) Briefly describe the structure of metals. (3)

8.2 Periodic Table: Period 3

Look at Period 3 of the Periodic Table and answer the following questions.
(a) (i) Which of the elements are classified as metals? (1)
  (ii) Which of the elements are classified as non-metals? (1)
(b) Give three properties of an element which help you to decide whether it is a metal or a non-metal. (3)
(c) Why does argon not form any compounds? (2)
(d) Give two reasons for believing that silicon is neither a true metal nor a true non-metal. (2)

8.3 Periodic Table: Group VII

Look at Group VII of the Periodic Table, then answer the questions below.
- **(a)** (i) State two pieces of chemical evidence which place chlorine, bromine and iodine in the same Group. (2)
  - (ii) Chlorine, bromine and iodine atoms all have different electronic structures, but all three have one common feature which is responsible for their chemical similarity. What is it? (1)
  - (iii) Would you expect the oxides of chlorine, bromine and iodine to be acidic or basic? Why? (2)
- **(b)** (i) Predict two physical properties of fluorine (F) and two of astatine (AT). (4)
  - (ii) At room temperature and pressure bromine has a physical property which it shares with only one other element. State this physical property, and name the other element. (2)
- **(c)** Chlorine and bromine combine together to form the compound BrCl. What sort of bond do you think exists between the atoms in this compound? Explain your reasoning. (2)

8.4 Lithium

Lithium is in Group I of the Periodic Table. It burns in oxygen to give lithium oxide (Li_2O). Lithium reacts with cold water to give hydrogen and an alkaline solution of lithium hydroxide (LiOH). Lithium chloride solution reacts with sodium carbonate solution to give a white precipitate of lithium carbonate. When lithium carbonate is heated, it decomposes into lithium oxide and carbon dioxide.
- **(a)** (i) State one way in which the reactions of lithium resembles those of sodium. (1)
  - (ii) State two ways in which the reactions of lithium resembles those of magnesium. (2)
- **(b)** What would you expect to see if carbon dioxide were passed into lithium hydroxide solution until no further change took place? (2)
- **(c)** (i) Write the equation for the action of heat on lithium carbonate. (1)
  - (ii) Suggest the equation for the action of heat on solid lithium hydroxide. (2)
- **(d)** Write the ionic equation for the reaction between lithium chloride solution and sodium carbonate solution. (1)
- **(e)** What products would you expect, **(i)** at the positive electrode and **(ii)** at the negative electrode, if a solution of lithium chloride were electrolysed using carbon electrodes? (2)
- **(f)** Which element, other than a Group II element, would you expect to resemble magnesium? (1)

8.5 Periodic Table: Calculation

Rubidium (Rb) is in the same Group of the Periodic Table as sodium and potassium. The commonest isotope of rubidium is $^{85}_{37}Rb$.
- **(a)**
  - **(i)** Write the formula of the rubidium ion. (1)
  - **(ii)** How many neurons are present in the nucleus of an atom of the commonest isotope of rubidium? (1)
- **(b)** Another isotope of rubidium has a mass number of 87.
  - **(i)** Write the symbol for this isotope. (1)
  - **(ii)** How are the atoms of this isotope different from those of $^{85}_{37}Rb$? (1)
- **(c)** Suggest how rubidium will react with water. Write an equation for the reaction. (4)
- **(d)** Rubidium reacts with oxygen according to the equation

  $$Rb(s) + O_2(g) \longrightarrow RbO_2(s)$$

  - **(i)** Calculate the maximum mass of rubidium oxide that could be made from 4.25 g of rubidium 85. (4)
  - **(ii)** What volume of oxygen, measured at r.t.p., would be used in this reaction? (2)
  - **(iii)** How many molecules of oxygen would this volume contain? (2)

9 The halogens

9.1 Chlorine

(a) Draw a labelled diagram of the apparatus and the chemicals you would use to prepare and collect reasonably pure, dry chlorine from concentrated hydrochloric acid (other than by electrolysis). Write the equation for the chemical reaction that occurs. (7)

(b) How does chlorine react with each of the following? In each case say what you would observe, and write the equation for each reaction.
  (i) sodium (3)
  (ii) hydrogen (3)
  (iii) dilute sodium hydroxide (2)
  (iv) methane (3)
  (v) ethene (2)

9.2 Sodium chloride

This question is about industrial uses of sodium chloride. Study the following reaction scheme, and then answer the questions below.

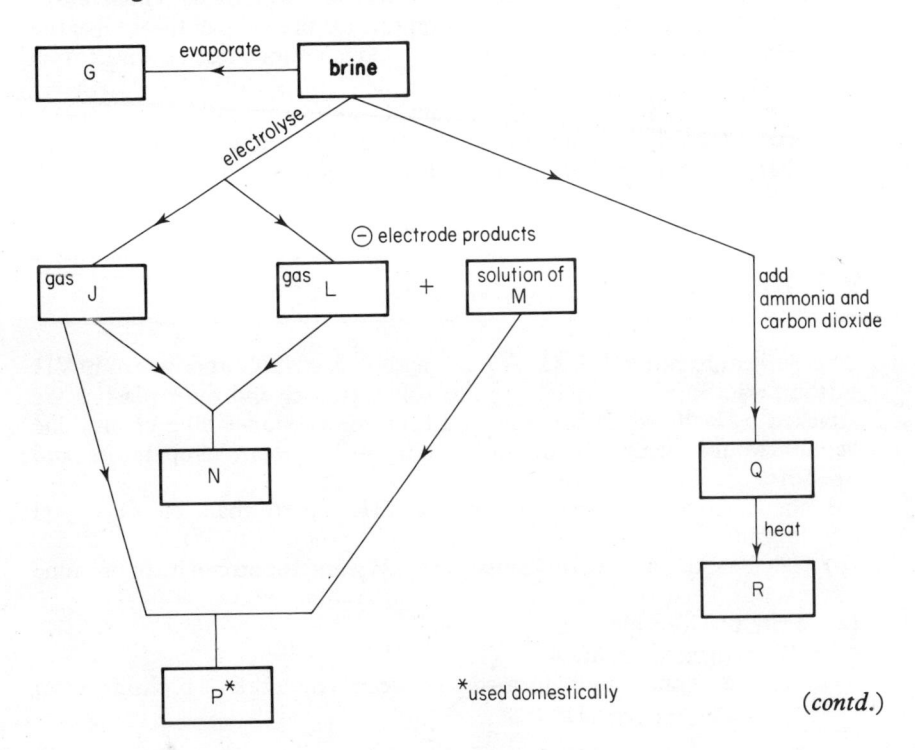

[Flow diagram: brine → evaporate → G; brine → electrolyse → ⊖ electrode products: gas J, gas L + solution of M; brine → add ammonia and carbon dioxide → Q → heat → R. J and L/M combine to N. N leads to P* (*used domestically).]

(contd.)

35

(a) Name each material which is represented by a letter. (8)
(b) How can gas J be obtained from sodium chloride other than by electrolysis? (3)

9.3 Fluorine

The element fluorine has atomic number 9 and mass number 19. At room temperature, fluorine is a gas and consists of diatomic molecules, F_2. Fluorine reacts vigorously with cold water, forming hydrogen fluoride and oxygen. It reacts with dilute sodium hydroxide to form fluorine monoxide.
(a) Give the number of protons, neutrons and electrons in an atom of fluorine. How are these electrons arranged around the nucleus? (4)
(b) Give the formulae of the compounds you would expect fluorine to form with
  (i) hydrogen (1)
  (ii) carbon (1)
  (iii) sodium. (1)
  Suggest the type of bonding, covalent or ionic, in each compound. (3)
(c) (i) Construct the equation for the reaction between fluorine and water. (2)
  (ii) The oxidation state of fluorine in fluoride monoxide is -1. What is the oxidation state of oxygen in this compound? (2)
(d) Fluorine is normally prepared by an electrolysis method. Which one of the following would be the most suitable for this preparation? Give the reason for your choice, and state why you have not chosen the other two methods.
  (i) Electrolysis of molten sodium fluoride
  (ii) Electrolysis of a solution of sodium fluoride in water
  (iii) Electrolysis of liquid hydrogen fluoride. (5)

9.4 Bromine

The element bromine melts at $-7.2°C$ and boils at $58°C$. It is in Group VII of the Periodic Table. It is very corrosive: rubber and most plastics are attacked by both liquid bromine and bromine vapour. 0.80 g of bromine vapour would occupy a volume of 120 cm^3 at room temperature and pressure.
(a) Name **two** members of Group VII, other than chlorine and bromine. (2)
(b) Calculate the relative molecular mass (M_r) and the atomicity of bromine vapour. (3)
(c) Write the formulae for
  (i) hydrogen bromide (1)
  (ii) the compound formed between hydrogen bromide and ammonia. (1)

(d) Bromine can be prepared by heating a mixture of sodium bromide, manganese(IV) oxide and concentrated sulphuric acid.
  (i) Draw a diagram of the apparatus you would use to carry out this preparation of bromine and to collect the liquid. (5)
  (ii) Suggest another compound that could be used instead of sodium bromide. (1)
(e) Bromine is obtained from sea-water, which contains sodium bromide, by passing chlorine through the sea-water. Write the ionic equation for this reaction. (1)

9.5 Oxidation and reduction

Study the following diagram:

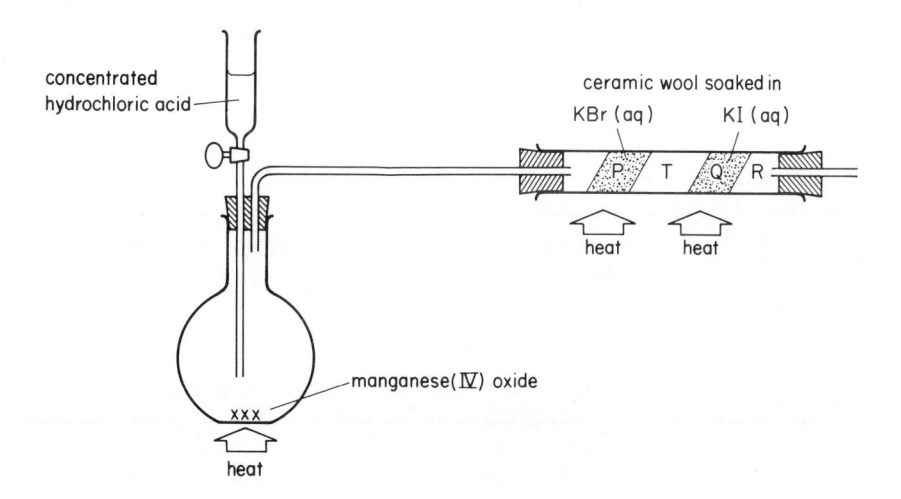

(a) (i) What gas is being prepared in the flask in this experiment? (1)
  (ii) Write the equation for the reaction between manganese(IV) oxide and concentrated hydrochloric acid. (1)
(b) (i) After a time, coloured products will be seen at P, Q, R and T. What will be the colours at P, Q, R and T? (4)
  (ii) Write equations to represent any reactions which have taken place in the combustion tube. (2)
(c) Which **three** chemicals in the experiment have acted as oxidising agents? (3)
(d) For **one** of these, state
  (i) what was oxidised by it (1)
  (ii) what were the products of the reaction. (2)

10 Nitrogen compounds

10.1 Ammonia

(a) Describe, with the aid of a diagram, how you would make a few jars of dry ammonia in the laboratory, starting from ammonium chloride.   (5)

(b) Describe and explain what you would see if
  (i) a jar of ammonia is inverted and opened under water   (2)
  (ii) a filter paper moistened with concentrated hydrochloric acid is brought near an open jar of ammonia   (3)
  (iii) a piece of damp red litmus paper is placed inside a jar of ammonia.   (2)

(c) Explain what is meant by 'thermal dissociation', and give an example of this process.   (4)

10.2 Nitric acid

(a) (i) Outline the manufacture of nitric acid from ammonia, giving essential conditions.   (7)
  (ii) Why is nitric acid manufactured from ammonia rather than directly from nitrogen?   (1)
  (iii) Give **two** industrial uses of nitric acid.   (2)

(b) 1.00 dm^3 of X, an oxide of nitrogen, exactly reacts with 2.67 g of hot copper metal to form copper(II) oxide and 1.00 dm^3 of nitrogen. (The volumes of both gases were measured at room temperature and pressure.) Deduce the equation for the reaction between X and copper.   (4)

10.3 Manufacture of ammonia

(a) Outline the manufacture of ammonia from nitrogen and hydrogen.   (6)

(b) Assuming that a 50% yield of ammonia is obtained in the commercial process, what mass of ammonia is obtained from 100 dm^3 of nitrogen and 300 dm^3 of hydrogen? Both volumes of gas are measured at room temperature and pressure.   (4)

(c) The reaction between nitrogen and hydrogen to make ammonia is exothermic. If the reaction to make ammonia is carried out at a low temperature, the speed of the reaction is much slower, but the yield of ammonia is greater. Why is this so?   (4)

10.4 Fertilisers and nitric acid

(a) Ammonium nitrate, NH_4NO_3, and ammonium sulphate, $(NH_4)_2SO_4$, are important fertilisers. Calculate the percentage by mass of nitrogen in each of these fertilisers and comment on the choice of one as a source of nitrogen. (3)

(b) Fertilisers are made from nitric acid. Nitric acid was first made commercially from sodium nitrate obtained from natural deposits of sodium nitrate. This method still forms the basis of the laboratory preparation of nitric acid. Draw a labelled diagram of the apparatus you would use. Name the other reagent used with sodium nitrate, and write the equation for the reaction. (7)

(c) Explain why using too much nitrogenous fertiliser may cause fish in rivers to die. (2)

10.5 Nitrogen dioxide

When dry lead(II) nitrate, $Pb(NO_3)_2$, is heated, and the products are passed through a U-tube surrounded by ice, a yellow liquid Y is formed in the U-tube and a colourless gas Z passes through. The yellow liquid contains 30.43 % of nitrogen and 69.57 % of oxygen by mass. Its relative molecular mass (M_r) is 92. The yellow liquid is very soluble in water, forming a mixture of nitric acid and nitrous acid, HNO_2.

(a) Calculate (i) the empirical formula and (ii) the molecular formula of the yellow liquid Y. (3)

(b) Name the colourless gas Z. (1)

(c) Suggest why lead(II) nitrate is used in this experiment and not copper(II) nitrate, $Cu(NO_3)_2, 3H_2O$. (2)

(d) What evidence is there in this experiment that
   (i) nitrogen is a non-metal
   (ii) the yellow liquid Y has a low melting point? (3)

(e) Write the equation for the reaction between the yellow liquid Y and aqueous sodium hydroxide. (2)

(f) How would you show that a solution of the yellow liquid Y in water contained nitrate ions? (3)

(g) When the yellow liquid Y is warmed, a brown gas, NO_2, is formed. When the brown gas is cooled in ice, the yellow liquid Y reforms. This can be represented by

yellow liquid Y $\rightleftharpoons$ brown gas

Write the chemical equation for this reaction, and explain the meaning of the sign $\rightleftharpoons$. (2)

11 Carbon compounds, fuels and hard water

11.1 Carbon dioxide

(a) Describe two different ways by which carbon dioxide can be obtained from calcium carbonate. (5)
(b) Describe two natural ways in which carbon dioxide is removed from the atmosphere. (5)
(c) Describe and explain what you would observe if magnesium were burnt in excess carbon dioxide. (3)
(d) When carbon dioxide is added to lime water the solution turns milky. Suggest two ways of removing this cloudiness, other than by filtration. (4)
(e) State two industrial processes that involve the use of carbon dioxide. (2)

11.2 Combustion of carbon

The diagram shows a section through a domestic fire, looked at from the side. The fuel being burnt consists of evenly sized lumps of a 'smokeless' material which can be thought of as carbon. The fire is burning well and the fuel is red hot.

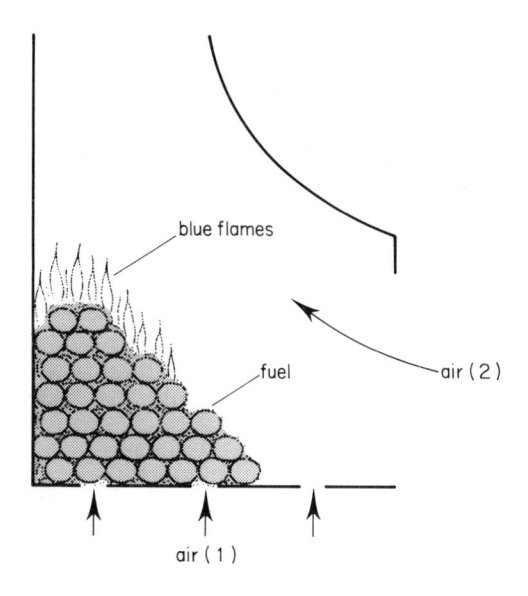

(a) Write the equation for the reaction which produces the majority of the heat thrown out by the fire.   (1)
(b) Whereabouts in the fuel does this take place?   (1)
(c) What causes the flickering blue flame on the top of the fire? Write the equation for the reaction taking place in the flames.   (3)
(d) Where was the substance which burnt with the blue flame produced in the fuel? Write the equation for the reaction by which it was produced.   (2)
(e) (i) Why does air from the room (labelled air (2)) flow over the fire?   (3)
   (ii) Why would it be dangerous to draught-proof the room too well?   (3)

11.3 Carbon monoxide

(a) Outline two industrial methods for preparing fuel gases which contain carbon monoxide.   (4)
(b) How and under what conditions does carbon monoxide react with
   (i) iron(III) oxide   (3)
   (ii) blood?   (3)
(c) Explain the presence of carbon monoxide in the exhaust gases from petrol engines.   (3)
(d) 10 cm^3 of carbon monoxide was exploded with 10 cm^3 of oxygen. Determine the volume and composition of the resulting gaseous mixture.   (4)

11.4 Limestone/hardness of water

Limestone rocks consist partly of calcium carbonate.
(a) Why is it that such rocks often contain large caves?   (5)
(b) Explain why water which comes from such areas causes kettles to 'fur up'.   (4)
(c) Why is this 'fur' a nuisance?   (2)
(d) How could the 'fur' be removed from the kettle?   (4)

11.5 Hardness of water 1

Four different samples of water were tested as shown below:

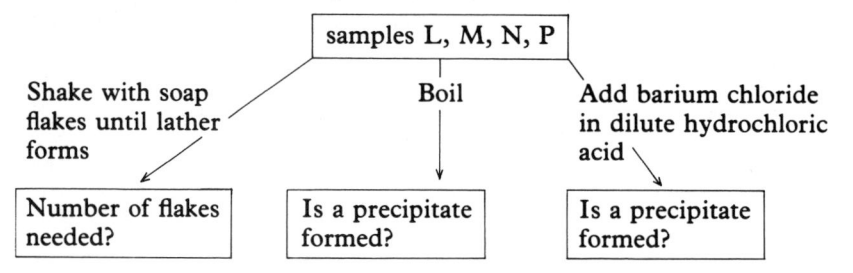

```
                    samples L, M, N, P
Shake with soap          Boil          Add barium chloride
flakes until lather                    in dilute hydrochloric
forms                                  acid
Number of flakes    Is a precipitate   Is a precipitate
needed?             formed?            formed?
```

The samples gave the following results:

Sample	Number of flakes needed to form a lather	Precipitate on boiling?	Precipitate with barium chloride?
L	1	no	no
M	4	yes, white	no
N	8	yes, white	yes, white
P	8	no	yes, white

(a) (i) What is meant by 'hard water'? (2)
(ii) Which sample was the softest? Explain. (2)
(iii) Which sample of water contains only permanent hardness? Explain. (2)
(iv) Which sample contains only temporary hardness? Explain. (2)
(b) Write an equation to represent the formation of the white precipitate made by boiling sample N. (1)
(c) Write an ionic equation to illustrate the formation of the white precipitate with barium chloride solution. (1)
(d) Name a method by which hard water can be softened in the home. Say briefly how the method works. (4)

11.6 Hardness of water 2

Temporary hard water can be softened either by boiling the water or by adding sodium carbonate.
(a) Name one substance responsible for causing temporary hardness in water. (1)

(b) Name two gases that you would expect to find in bubbles from boiling temporarily hard water. (2)
(c) State (i) one advantage of hard water (1)
  (ii) one disadvantage of hard water. (1)
(d) Explain how the addition of sodium carbonate removes the temporary hardness in water. (2)
(e) The ability of temporarily hard water to conduct electricity falls when the water is boiled. When just sufficient sodium carbonate is added to remove the hardness from the same sample of water, the ability to conduct electricity hardly changes. Explain these observations. (4)

12 Carbon chemistry

12.1 Crude oil

Crude oil is a mixture of many compounds.
(a) Which two chemical elements make up most of these compounds? (2)
(b) (i) Name the process by which crude oil is refined. (1)
  (ii) Draw a labelled diagram of a laboratory apparatus which demonstrates this type of separation. (6)
(c) CH_4 and C_4H_{10} are the formulae of two compounds which may be extracted from crude oil. Give the name of each and say what it is commonly used for. (4)

12.2 Alkanes

Propanal, C_2H_5CHO, and two saturated hydrocarbons X and Y have the same relative molecular mass. Propanal burns to give carbon dioxide and water only.
(a) What is meant by (i) saturated, (ii) hydrocarbon, (iii) isomers? (5)
(b) (i) Deduce the molecular formula of X and Y. (2)
  (ii) Write down the full structural formulae of X and Y. (2)
(c) Write equations for the complete combustion of (i) X and (ii) propanal. (2)
(d) Propanal, boiling point 49°C, is made by slowly adding a solution of sodium dichromate(VI) and propanol to hot dilute sulphuric acid in a distillation flask. The dichromate(VI) and alcohol mixture are added to the acid at the same rate as the propanal distils over.
  (i) Draw a suitable apparatus for the preparation and collection of propanal. (4)

(contd.)

(ii) Suggest two liquid impurities that might be present in propanal made by this method. (2)

12.3 Homologous series

A, B, C and D are four homologous series of organic compounds whose general formulae can be represented as shown below:
A C_nH_{2n+2}
B C_nH_{2n}
C $C_nH_{2n+1}OH$
D $C_nH_{2n+1}CO_2H$
(a) Name the homologous series represented by A, B and C. (3)
(b) Give the name and the structural formula of a typical member from (i) series A, (ii) series B. (2)
(c) What is the name of the type of reaction that takes place between a compound from series C and a compound from series D? (1)
(d) How can
 (i) a member of series B be obtained from a member of series C (3)
 (ii) a member of series D be obtained from a member of series C? (3)
(e) When 1 mole of an organic compound was burnt in oxygen, 3 moles of carbon dioxide and three moles of water were formed. Assuming that the organic compound belongs to one of the four series A, B, C or D, deduce what you can about the compound. (3)

12.4 Methane and ethane

Sodium ethanoate, CH_3CO_2Na, can be used, under suitable conditions, to prepare both methane and ethane. Methane is obtained by heating a mixture of solid sodium ethanoate with solid sodium hydroxide. Ethane is obtained, together with carbon dioxide, at the anode (positive electrode), when aqueous sodium ethanoate is electrolysed.
(a)   (i) What is the main natural source of methane? (1)
       (ii) Methane and ethane are in the same homologous series. Name this series, and give three general properties of homologous series. (4)
       (iii) Methane, under suitable conditions, reacts with steam to give carbon monoxide and hydrogen only. Construct the equation for this reaction. (2)
(b)   (i) Draw a diagram to show how you would prepare methane from solid sodium ethanoate and solid sodium hydroxide. (3)
       (ii) Write down the formulae of the two ions present in sodium ethanoate. Hence explain the formation of ethane and carbon dioxide at the anode when aqueous sodium ethanoate is electrolysed. (4)

(iii) How would you separate ethane from a mixture of ethane and carbon dioxide? (1)
(iv) What substances would you expect to be formed at the anode if aqueous sodium methanoate were electrolysed instead of sodium ethanoate? (2)

12.5 Alkenes

(a) Briefly describe how ethene can be obtained from ethanol. (3)
(b) (i) Describe what you would expect to observe when ethene is reacted with bromine. (2)
   (ii) Write down the full structural formula of the product. (1)
   (iii) Write down the empirical formula of the product. (1)
   (iv) Write down the full structural formula of the isomer of the product. (1)
(c) Tetraethyl-lead(IV) is added to petrol to make the engine run smoothly. The compound obtained in (b)(i) is also added to petrol, to ensure that lead escapes in the exhaust fumes, instead of being deposited in the engine.
   (i) Suggest the name of the lead compound present in the exhaust fumes. (1)
   (ii) What are the dangers of lead pollution? (2)
(d) Carbon monoxide is also present in car exhaust fumes. Describe how the carbon monoxide is formed, and why it is harmful to health. (4)
(e) The sparking plugs in car engines cause nitrogen and oxygen in the air to react and form nitrogen monoxide. Nitrogen monoxide reacts with the air to form nitrogen dioxide. What are the dangers of nitrogen dioxide in the air? (2)

12.6 Organic structures

(a) The diagrams show six full structural formulae. Beside them is a list of names. Which name corresponds to which formulae?

1.
```
    H  Cl
    |  |
H—C—C—Cl
    |  |
    H  H
```

2.
```
 Cl       Cl
   \    /
    C=C
   /    \
  H       H
```

1,1-dichloroethane
hexachloroethane
chloroethene
1,2-dichloroethane
1,1,2-trichloroethene
1,2-dichloroethene

3.
```
  Cl      Cl
    \   /
     C=C
    /   \
  Cl      H
```

4.
```
    Cl  Cl
    |   |
Cl—C—C—Cl
    |   |
    Cl  Cl
```

(contd.)

5
$$\begin{array}{c} H \\ \diagdown \\ C=C \\ \diagup \\ H \end{array} \begin{array}{c} H \\ \diagup \\ \diagdown \\ Cl \end{array}$$

6
$$\begin{array}{c} Cl \; Cl \\ | \;\; | \\ H-C-C-H \\ | \;\; | \\ H \; H \end{array}$$

(b) (i) Which one of these compounds, when polymerised, would produce poly(vinylchloride), PVC, whose structural formula is

$$-\overset{H}{\underset{H}{C}}-\overset{Cl}{\underset{H}{C}}-\overset{H}{\underset{H}{C}}-\overset{Cl}{\underset{H}{C}}-\overset{H}{\underset{H}{C}}-\overset{Cl}{\underset{H}{C}}-\overset{H}{\underset{H}{C}}-\overset{Cl}{\underset{H}{C}}-\overset{H}{\underset{H}{C}}-\overset{Cl}{\underset{H}{C}}- \; ?$$
(1)

(ii) State and explain one physical difference and one chemical difference between PVC and the monomer from which it was formed. (4)

(iii) State two uses of PVC and explain why it has these uses. (4)

12.7 Monomers and polymers

(a) What is meant by a **monomer**? Give two **physical** differences you would expect there to be between a monomer and its polymer. (6)

(b) A sample of poly(ethene) has an average relative molecular mass of 28 000. How many ethene molecules were combined into the average polymer molecule? (2)

(c) Give one use for poly(ethene) and give two reasons which make the polymer suitable for that use. (3)

(d) What is meant by photosynthesis? What polymerisation process takes place in photosynthesis? (5)

12.8 Proteins

A piece of cheese contains protein. Proteins are polymers of amino acids, which contain the element nitrogen. This nitrogen may have originated from the air and its course can be traced as shown in the following flow diagram:

air ⟶ soil ⟶ grass ⟶ cow ⟶ cheese

(a) At which of the four stages are amino acids first formed? (1)

(b) Briefly describe what happens to the amino acids from this stage to the cheese product. (2)

(c) By what natural chemical reactions is the nitrogen in the air converted to a form suitable for absorption by plants? (4)

(d) An element present in bread is carbon. The course of carbon can be traced as shown in the following flow diagram:

carbon dioxide in air ⟶ glucose ⟶ starch ⟶ flour ⟶ bread

(i) What is the name of the process by which carbon dioxide becomes glucose?   (1)
   (ii) What else is necessary, besides carbon dioxide, for the process **(d) (i)** to take place?   (3)
(e) What is the chemical relationship between glucose and starch?   (1)
(f) Give one way of converting starch into glucose.   (2)
(g) By what processes is the amount of carbon dioxide in the air kept constant?   (4)
(h) What compound might you add to the atmosphere in a greenhouse to increase plant growth?   (1)

12.9 Hydrocarbons/heats of combustion

The table below gives the names and structures of various hydrocarbons together with their heats of combustion. Heat of combustion is the heat change when 1 mole of a substance is completely burnt in oxygen. The value is always negative, showing that combustion reactions are exothermic.

Substance	Structural formula	Heat of combustion /kJ mol^{-1}
methane	H | H—C—H | H	−890
ethane	H H | | H—C—C—H | | H H	−1560
ethene	H H \\ / C=C / \\ H H	−1409
propane	H H H | | | H—C—C—C—H | | | H H H	−2230

(a) What is meant by the term 'hydrocarbon'?   (1)
(b) By reference to ethane and ethene, explain the meanings of the following terms
   (i) substitution   (2)
   (ii) addition.   (2)

(contd.)

(c) Write equations for the complete combustion of
   (i) ethane   (1)
   (ii) ethene.   (1)
(d) Suggest a reason why the heat of combustion of ethene is less than that of ethane.   (1)
(e) Predict, with reasons, the heats of combustion of
   (i) butane   (2)
   (ii) ethyne, $H-C \equiv C-H$.   (2)

13 Metals and the reactivity series

13.1 Reactivity series

(a) List the following metals in the order of the reactivity series, with the most reactive first:

copper   iron   magnesium   silver   sodium   zinc   (3)

(b) The diagrams below show pieces of metal dipping into solutions of metal salts.

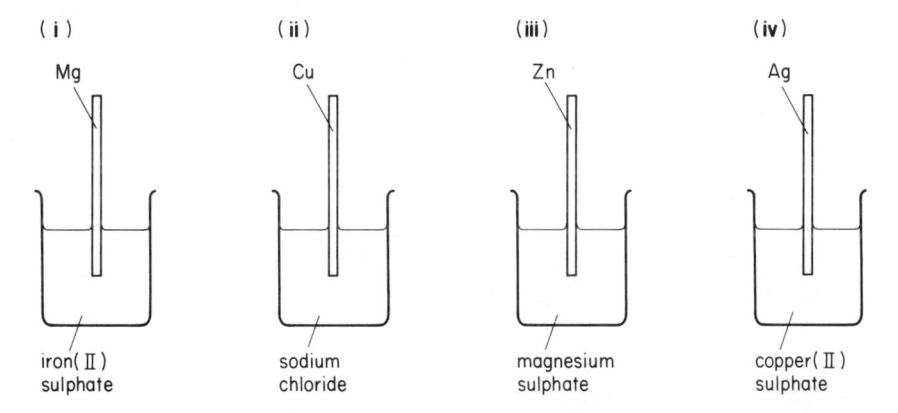

(i) Mg in iron(II) sulphate

(ii) Cu in sodium chloride

(iii) Zn in magnesium sulphate

(iv) Ag in copper(II) sulphate

For each of the examples **(i)** to **(iv)**, state whether any reaction will take place. Where a reaction does take place, write an ionic equation to represent the change, and state which particles are oxidised and which particles are reduced.   (10)

13.2 Transition metals

The order of reactivity of four transition metals (most reactive first) is: manganese, iron, nickel, (hydrogen), copper.

(a) A typical property of transition metals is that the metals or their compounds are catalysts. Give an example of each of these metals, or their compounds, acting as a catalyst. (No equations are required.)   (4)

(b) Give two other typical properties of transition metals or their compounds.   (2)

(c) Zinc is sometimes included in the transition metals. Give two properties of zinc or its compounds which are not typical of transition metals.   (2)

(d) Suggest how nickel can be obtained from nickel(II) oxide (other than by electrolysis), giving a reason for your answer.   (2)

13.3 Manufacture of iron

The diagram below shows the blast furnace for the manufacture of iron.

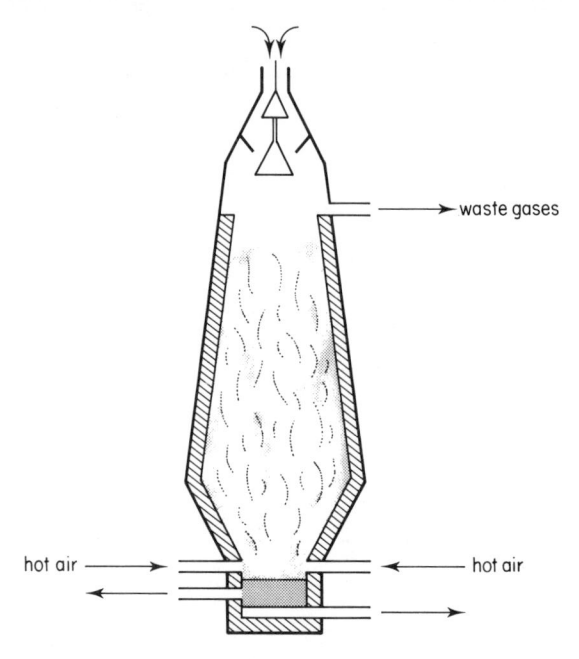

(a) Explain why the waste gases contain
  (i) carbon dioxide   (5)
  (ii) carbon monoxide   (3)
  (iii) nitrogen.   (1)
(b) If the iron(III) oxide is wet, the waste gases also contain hydrogen. Explain how hydrogen is formed, and why the formation of this gas is dangerous.   (3)
(c) Suggest a suitable refractory oxide for the brickwork.   (1)
(d) How is sand removed from the furnace? Why is it necessary to remove the sand?   (2)
(e) **(i)** Name the element present in pig iron that makes it hard but brittle.   (1)
  (ii) Explain how this element is removed by blowing air/oxygen through the molten pig iron.   (2)
(f) Iron objects rust easily. Give two different ways of preventing rusting.   (2)

13.4 Copper compounds

A mixture of copper(II) hydroxide and copper(II) carbonate can occur as either malachite, $Cu(OH)_2$, $CuCO_3$, or azurite, $Cu(OH)_2$, $2CuCO_3$.

(a) Describe what you would expect to see when either malachite or azurite is
   (i) heated strongly   (2)
   (ii) treated with dilute sulphuric acid.   (2)
(b) Describe an experiment you would carry out in order to find the percentage loss in mass when either malachite or azurite is heated in air until no further change occurs.   (6)
(c) A compound is known to be either malachite or azurite. If the percentage loss in mass in (b) is found to be 28.0%, prove the compound must be malachite. Explain your reasoning.   (4)

13.5 Tests for metals

A chemist was given four pieces of metal. He was told that one of them was aluminium, one was copper, one was iron and one was sodium. He was asked to discover which was which. He decided to use three tests. The tests are shown in the reaction scheme.

Treat the metal with cold water

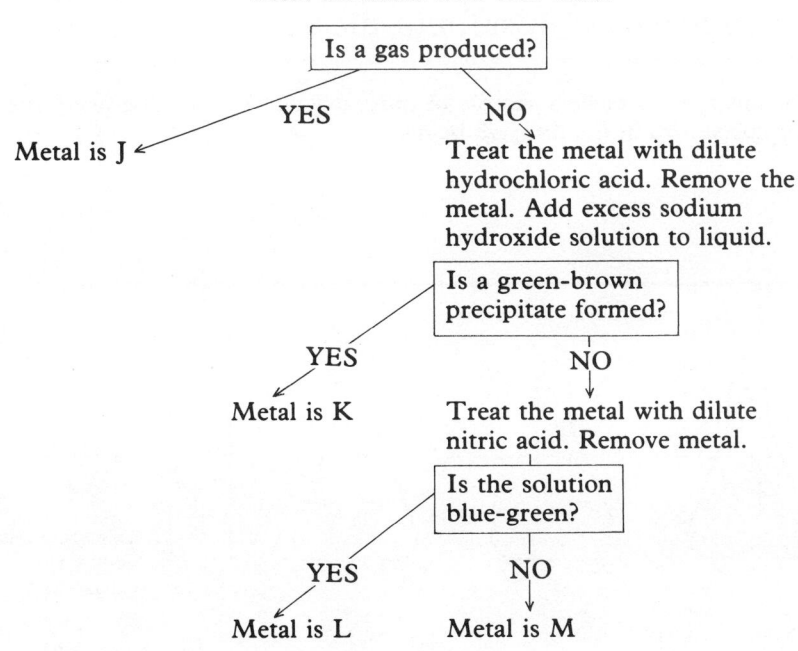

Is a gas produced?

YES → Metal is J

NO → Treat the metal with dilute hydrochloric acid. Remove the metal. Add excess sodium hydroxide solution to liquid.

Is a green-brown precipitate formed?

YES → Metal is K

NO → Treat the metal with dilute nitric acid. Remove metal.

Is the solution blue-green?

YES → Metal is L

NO → Metal is M

(a) Identify each of the metals J, K, L and M.   (4)
(b) Explain what has happened in each of the following tests. Write equations for any reactions that take place.
   (i) Metal J treated with water.   (2)
   (ii) Metal K treated with hydrochloric acid, then sodium hydroxide solution added.   (4)
   (iii) Metal L treated with dilute nitric acid.   (2)

14 Sulphur and sulphur compounds

14.1 Sulphur

You are provided with a finely powdered mixture of sulphur and lead(II) oxide.
(a) Describe how you would prepare from this mixture
    (i) a crystalline sample of pure sulphur   (4)
    (ii) a pure sample of an insoluble salt of lead   (6)
(b) How could you prepare from the crystalline sulphur in (a) (i) a sample of a different crystalline form of sulphur?   (2)
(c) Name the two allotropic forms of carbon.   (1)
(d) State and explain what you would observe if a mixture of carbon and lead(II) oxide were heated.   (4)

14.2 Preparation of sulphur dioxide

A student tries to make a sample of pure, dry sulphur dioxide using the apparatus shown in the diagram below.

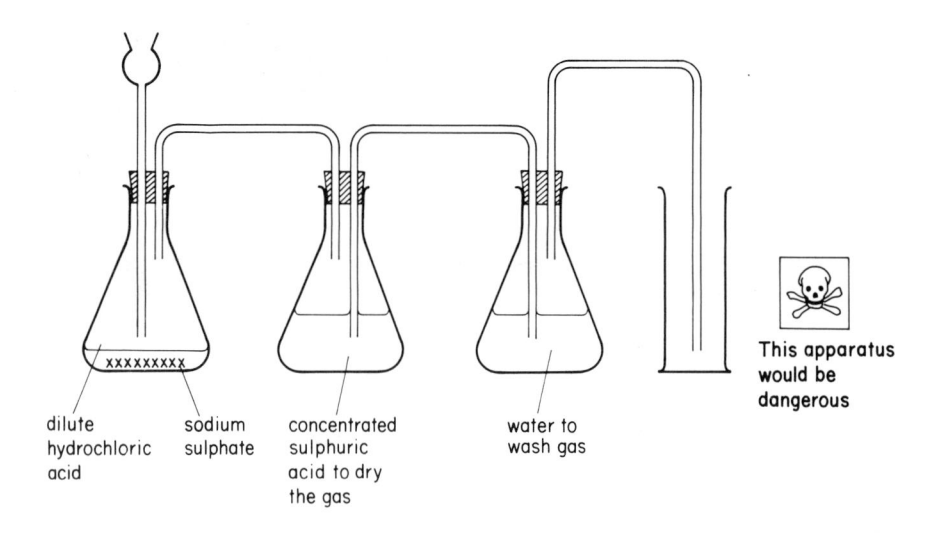

(a) List **six** things which are wrong with this apparatus.   (6)
(b) Explain why the apparatus as set up is **dangerous**.   (4)
(c) Redraw the diagram to show how the apparatus should be used.   (5)

14.3 Sulphuric acid

(a) Describe in outline the manufacture of sulphuric acid from sulphur. (7)
(b) Sulphur dioxide reduces a solution containing Fe^{3+} ions to Fe^{2+} ions. When aqueous sodium hydroxide is added to the resulting solution drop by drop, no precipitate appears until several drops have been added.
  (i) Name the sulphur-containing compound formed in the reaction between Fe^{3+} ions and sulphur dioxide. (1)
  (ii) Name the precipitate formed on adding aqueous sodium hydroxide. (1)
  (iii) Explain why the precipitate does not appear until several drops of aqueous sodium hydroxide have been added. (1)
(c) Describe and explain the difference in reactions when sulphur dioxide is bubbled into
  (i) barium chloride solution (2)
  (ii) barium chloride solution acidified with dilute hydrochloric acid. (2)

14.4 An industrial process

The flow diagram below represents the industrial process by which a certain common acid X is made.

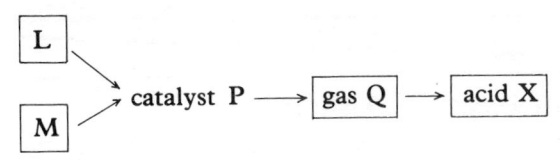

```
L ⟶
      catalyst P ⟶ gas Q ⟶ acid X
M ⟶
```

(a) (i) Name the substances L, M, P, Q and X. (5)
    (ii) Give the formula of the acid X. (1)
(b) In practice, the gas Q is not added directly to water. Why not? How is the conversion of Q to X carried out? (4)
(c) (i) What mass of X is needed to react completely with 5.0 g of sodium hydroxide? (4)
    (ii) Name the product which is formed if only half the mass of acid is reacted with the same mass of sodium hydroxide. (1)

14.5 Sulphates

When a green crystalline compound was heated, it first lost its water of crystallisation, and the original green crystalline solid was converted to a yellow anhydrous solid. *(contd.)*

When 3.04 g of the anhydrous solid was heated, it gave 240 cm³ of sulphur dioxide and 240 cm³ of sulphur trioxide (both volumes corrected to room temperature and pressure). 1.60 g of the reddish-brown solid Fe_2O_3 was left.
- (a) Give one chemical test each for (i) water and (ii) sulphur dioxide. (4)
- (b) Draw a diagram of the apparatus you would use to obtain a pure, dry sample of sulphur dioxide from the gases given off when the yellow solid is heated. (3)
- (c) (i) Calculate the mass in grams of the anhydrous yellow compound required to produce 24 dm³ of sulphur dioxide at room temperature and pressure. (1)
  - (ii) If the relative formula mass of the anhydrous yellow compound is 152, calculate the mass of iron(III) oxide, and the volume of sulphur trioxide (measured at room temperature and pressure), when 1 mole of the anhydrous yellow compound is heated. (2)
  - (iii) Hence construct the equation for the action of heat on the yellow compound. (4)

15 Air, oxygen, hydrogen and redox reactions

15.1 Percentage of oxygen in the air

The percentage of oxygen in the air can be measured using the apparatus shown in the diagram below.

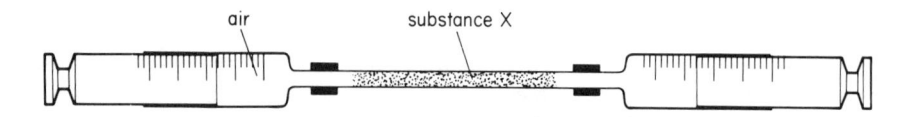

- (a) (i) Name a suitable metal for X. (1)
  - (ii) Briefly describe how you would perform this experiment. (6)
- (b) The percentage by volume of oxygen in the air obtained from boiling out the air dissolved in water, was found to be 33%. Explain why this result is different from the value for the percentage of oxygen found in the air. (2)

(c) Describe and explain the effects of exposing each of the following to air:
   (i) iron (2)
   (ii) aqueous calcium hydroxide (2)
   (iii) hydrated sodium carbonate (2)
   (iv) anhydrous calcium chloride (2)
   (v) aqueous iron(II) sulphate. (2)

15.2 Oxygen 1

(a) Describe, with the aid of a diagram, how you would prepare a sample of pure, dry oxygen in the laboratory. (7)

(b) Describe what is seen when each of the following are introduced separately into jars of oxygen. (No equations are required.)
   (i) a glowing splint (1)
   (ii) a piece of damp blue litmus paper (1)
   (iii) some burning sulphur (1)
   (iv) red-hot iron wire (2)
   (v) a small piece of white phosphorus. (2)

(c) What volume of oxygen, measured at room temperature and pressure, would be needed to combine completely with 2.4 g of magnesium? (4)

15.3 Air

A student set up the apparatus shown in the diagram below.

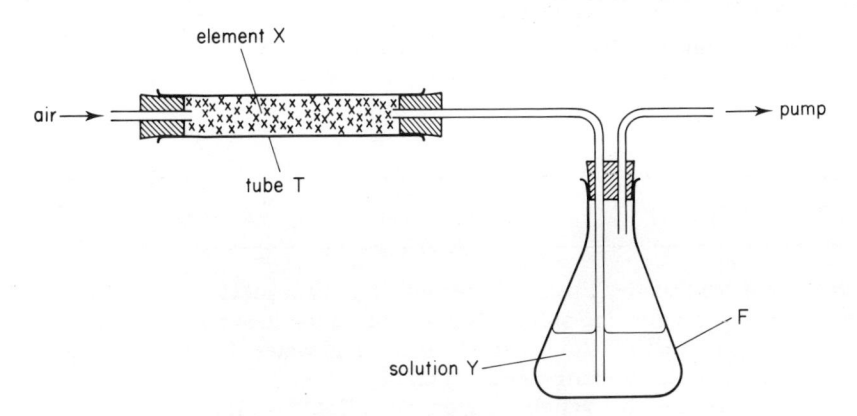

The tube T is filled with an element X, and the flask F contains a colourless solution Y.

The student makes the following observations:
   (I) If air is drawn slowly through the apparatus for some time, a white precipitate forms in the flask F. (*contd.*)

(II) If the tube T is heated strongly and the air continues to flow, more white precipitate quickly forms in flask F and then slowly disappears.
(III) If the flask F is now removed and heated until the liquid in it boils, the white precipitate comes back again.
(a) Identify X and Y. (2)
(b) (i) Explain the observations in I. (3)
   (ii) Explain the observations in II. (6)
   (iii) Explain the observations in III. (3)

15.4 Oxides

Read carefully through the following list of oxides:

aluminium oxide	calcium oxide	carbon dioxide
carbon monoxide	copper(II) oxide	hydrogen oxide (water)
magnesium oxide	nitrogen dioxide	sodium oxide
sulphur(VI) oxide	sulphur dioxide	zinc oxide

From this list
(a) Make a list of those oxides which are acidic but not amphoteric. (3)
(b) Make a list of those oxides which are amphoteric. (1)
(c) Make a list of those oxides which are basic but not amphoteric. (3)
(d) Make a list of those oxides which are neutral. (2)
(e) Briefly describe how you would attempt to prove experimentally that you have put
   (i) zinc oxide
   (ii) sulphur dioxide
   into the right class of oxide. (6)
(f) Write the molecular equation for the reactions between
   (i) copper(II) oxide and dilute sulphuric acid (1)
   (ii) sodium hydroxide and dilute hydrochloric acid. (1)
(g) Write the ionic equations for (f)(i) and (f)(ii). (2)

15.5 Oxygen 2

When red lead oxide, Pb_3O_4, is heated, it gives lead(II) oxide and a gas which relights a glowing splint. When red lead oxide is reacted with nitric acid it forms lead(IV) oxide, lead(II) nitrate and water. Lead(IV) oxide has similar properties to manganese(IV) oxide.
(a) Name the gas that relights a glowing splint. (1)
(b) What gas would be formed when lead(IV) oxide is added to hydrogen peroxide? (1)
(c) Construct the equations for
   (i) the action of heat on red lead oxide (1)
   (ii) the reaction of red lead oxide with dilute nitric acid. (1)
(d) (i) Name the three products formed when red lead oxide reacts with

hot concentrated hydrochloric acid. Write the question for this reaction. (4)
  (ii) Why does this reaction give a white precipitate when cooled to room temperature? (2)
(e) Red lead oxide is a mixture of two oxides of lead. Suggest the formulae and the mole ratio of these two oxides. (2)

15.6 Acidic and basic oxides

(a) Write down formulae for one example of each of the following oxides, stating whether each oxide contains ions or molecules.
  (i) an acidic oxide   (1)
  (ii) a basic oxide   (1)
  (iii) an oxide that can be reduced when heated in hydrogen.   (1)
(b) The metal molybdenum, relative atomic mass (A_r) 96, forms two basic oxides of formulae Mo_2O and Mo_2O_3.
  (i) Write down, for each of the oxides, the mass of oxygen that combines with one mole of molybdenum. (2)
  (ii) Write the formulae of the ions present in solution when each of the oxides Mo_2O and Mo_2O_3 react with hydrochloric acid. (2)
  (iii) What volume of 1.0 mol/dm^3 hydrochloric acid will be neutralised by 1 mole of the oxide Mo_2O_3? (3)
  (iv) Suggest, in outline, a method by which a pure sample of Mo_2O_3 might be prepared from the solution obtained in (b)(iii). (5)

15.7 Oxidation and reduction

(a) Define oxidation in terms of
  (i) electron transfer   (1)
  (ii) change in oxidation state (number).   (1)
(b) (i) Name a gas, other than ozone, which contains oxygen and can act as an oxidising agent.   (1)
  (ii) Name a gaseous oxidising agent that does not contain oxygen.   (1)
  In each case give an example of the gas acting as an oxidising agent.   (2)
(c) In each of the following reactions name the oxidising agent and state the oxidation process taking place.
  (i) $Zn(s) + NiSO_4(aq) \longrightarrow Ni(s) + ZnSO_4(aq)$ (2)
  (ii) $MnO_2(s) + 4HCl(aq) \longrightarrow MnCl_2(aq) + Cl_2(aq) + 2H_2O(l)$ (2)
(d) Most metals are extracted by a reduction process. Give, with one example in each case, two different types of reduction process used in the extraction of metals.   (4)

15.8 Hydrogen as a reducing agent

2.00 g of copper(II) oxide is reduced using the apparatus shown in the diagram below.

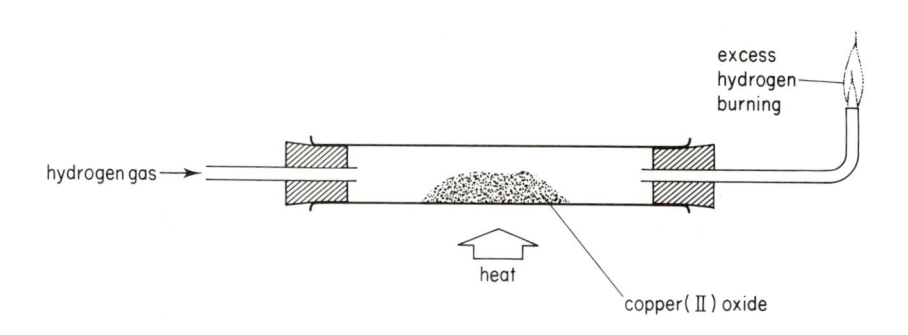

Initially, hydrogen gas is allowed to pass through the tube for about 15 seconds, before it is lighted at its outlet. The copper(II) oxide is then heated strongly. After the reaction is completed, a small supply of hydrogen gas is continued while the copper formed in the reaction is allowed to cool.

(a) Write the equation for the reaction between copper(II) oxide and hydrogen.   (1)
(b) Why must the hydrogen be lighted?   (2)
(c) State three observations you might make while the reaction is carried out.   (3)
(d) Give two ways by which you could tell that the reaction had gone to completion.   (2)
(e) Why must a small supply of hydrogen gas be continued while allowing the copper to cool?   (2)
(f) What is the maximum mass of copper that could be produced in this experiment?   (3)
(g) Explain whether similar experiments could be carried out
   (i) using carbon monoxide in place of hydrogen
   (ii) using magnesium oxide in place of copper(II) oxide.
   Give reasons for your answers.   (2)

16 Technological, social, economic and environmental chemistry

16.1 Manufacturing processes

The terms below are used in manufacturing processes.

alloy	ash	by-product	chalk
detergent	fuel	ore	raw materials
pollutant	refine	refractory	roast
rock	salt	slag	soap
steel	yield		

Choose a term from this list to match each of the definitions below.
(a) The waste product, including calcium silicate, obtained from a blast furnace. (1)
(b) A substance that is burned to give heat energy, e.g. the coke used in a blast furnace. (1)
(c) A mineral that is a compound from which a metal may be profitably extracted, e.g. aluminium from bauxite. (1)
(d) A mixture of two or more metals, or of a metal and a non-metal, e.g. brass. (1)
(e) The amount of product obtained from an industrial process, described as a percentage of the amount which the reactants could produce, e.g. 10% for nitrogen and hydrogen reacting in the Haber Process to make ammonia. (1)
(f) A salt of a Group I metal and a carboxylic acid, e.g. potassium stearate. (1)
(g) A substance produced in the manufacture of something else, e.g. chlorine in the manufacture of sodium hydroxide. (1)
(h) The process by which impurities are removed from a substance, e.g. to obtain petrol from crude oil. (1)
(i) The substances out of which a chemical process makes its finished products, e.g. limestone, sodium chloride and ammonia for the manufacture of sodium carbonate. (1)
(j) Describe a substance that can be heated to a high temperature without decomposing e.g. bricks. (1)

16.2 Discovery of the elements

The table below gives the symbol for the element, the date of discovery, and the name of the discoverer.

H	Cavendish	1766	He	Lockyer & Jannsen	1868
Li	Arfwedson	1817	Be	Vaquelin	1798
B	Gay-Lussac & Davy	1808	C	known in ancient times	
N	Rutherford	1772	O	Scheele & Priestley	1774
F	Scheele	1771	Ne	Ramsay & Travers	1898
Na	Davy	1807	Mg	Davy	1808
Al	Oersted	1825	Si	Berzelius	1824
P	Brand	1669	S	known in ancient times	
Cl	Scheele	1774	Ar	Rayleigh & Ramsay	1894
K	Davy	1807	Ca	Davy	1808
Np	Seaborg	1940	Pu	Seaborg	1940

(a) Give four examples of pairs of elements in the same Group of the Periodic Table that were discovered by the same scientist in approximately the same year. (4)

(b) (i) Explain why elements such as carbon and sulphur were known in ancient times. (1)

(ii) Name two other elements, not in the table, that were known in ancient times. (2)

(c) In 1789, Lavoisier listed the chemical elements. Suggest why he thought compounds such as calcium oxide, aluminium oxide and silicon dioxide were elements. (1)

(d) In 1800 it became possible to produce a continuous supply of an electric current. Which of the elements in the table do you think were discovered by electrolysis? (3)

(e) Silicon was discovered by reacting silicon dioxide with magnesium. Suggest another element that could have been discovered by reacting its oxide with magnesium. (1)

(f) Suggest why neon and argon were not discovered until after 1890. (1)

(g) Suggest when each of the following elements P, Q, R and T might have been discovered.

(i) Element P is monatomic and is very unreactive. (1)

(ii) Element Q combines in the ratio one mole of the element to one mole of oxygen. The element reacts with water to give hydrogen. (1)

(iii) Element R does not react with water and forms an important alloy with zinc. (1)

(iv) Element T is radioactive. It is made by bombarding plutonium with alpha particles. (1)

16.3 Modern materials

Over the years plastics have come to be used as alternatives to many other materials. Use the information in the table to answer the questions.

Object	Old material	New material
Windows	glass	Perspex
Drain pipes	iron	PVC
Storage boxes	aluminium	polystyrene
Water pipes	copper	polypropene
Chemical stoppers	cork	polythene

(a) (i) Give one advantage that Perspex has over glass. (1)
    (ii) Give one advantage that glass has over Perspex. (1)
(b) Give two advantages that PVC (polyvinyl chloride) has over iron. (2)
(c) Suggest the two poisonous gases that are formed when PVC burns. (2)
(d) Suggest reasons for the following:
    (i) aluminium is suitable for cooking foil (2)
    (ii) plastics are unsuitable for cooking foil. (2)
(e) Give a use of copper which will not be replaced by plastics. (1)
(f) Give one advantage polythene has over cork for use in stoppers. (1)
(g) Name two plastics that are used as materials for clothing. (2)
(h) Name
    (i) a chemical that will react with iron but not with a plastic (1)
    (ii) a chemical that will affect plastic but not iron. (1)

16.4 Disposal of household waste

Household waste contains paper, plastics (such as polythene and polyvinyl chloride), glass, wood and metals (such as iron and aluminium). Many of these materials could be reclaimed and then reused. Most waste is got rid of by either burning or burying.

Waste is burnt in metal incinerators using an excess of air to ensure complete oxidation. Compounds such as carbon dioxide and water are produced. One of the problems from this method of disposal is the formation of hydrogen chloride.

In areas where air pollution is a problem, waste is usually buried. Waste is tipped to a depth of about 10 metres and then compacted with bulldozers.

(contd.)

The waste is then covered with soil. This process is repeated until the hole has been completely filled. Over a period of time, the organic matter decomposes by the action of anaerobic bacteria, producing methane and carbon dioxide. The problem of methane accumulation and subsidence makes it impossible to build houses on or near these sites. There is also a water pollution problem in areas of large rainfall.

(a) Name
   (i) three different types of waste materials that will burn to form carbon dioxide and water.   (3)
   (ii) two different types of waste materials that will not burn.   (2)
(b) What harmful gas would be formed if a limited supply of air were used?   (1)
(c) (i) Which waste material gives off hydrogen chloride?   (1)
   (ii) Give two disadvantages caused by the formation of hydrogen chloride.   (2)
(d) Which material when buried would take
   (i) the shortest time to decompose?   (1)
   (ii) the longest time to decompose?   (1)
(e) Explain why iron corrodes more quickly than aluminium.   (2)
(f) Anaerobic bacteria act in the absence of oxygen. Give an everyday manufacturing process that uses anaerobic bacteria.   (1)
(g) Why is the accumulation of methane dangerous?   (1)
(h) How does the burying of waste cause water pollution?   (2)
(i) Suggest one method by which substances made of iron could be removed from waste material before it is burned or buried.   (1)
(j) Suggest two types of areas which would have high levels of air pollution, and two possible air pollutants found in these areas.   (4)

16.5 Limestone, lime and slaked lime

Calcium carbonate occurs naturally as chalk, limestone and marble. It is insoluble in water. When calcium carbonate is heated it decomposes to form calcium oxide. Calcium carbonate is an important raw material. It is used in the manufacture of iron and sodium carbonate.

Calcium oxide can be slaked by the addition of water to form an alkaline solution, slaked lime (calcium hydroxide). This reaction is exothermic. Calcium hydroxide is used in industry for the removal of acidic gases. In the laboratory a solution of calcium hydroxide in water is used as a test for carbon dioxide. Calcium hydroxide or calcium oxide is added to soil by farmers. Calcium hydroxide is also used to soften water in reservoirs. It removes temporary hardness, e.g. hardness caused by calcium hydrogencarbonate, but not permanent hardness, e.g. hardness caused by calcium sulphate. The exact amount of calcium hydroxide must be added.

(a) What are the chemical names for (i) limestone, (ii) slaked lime?   (2)
(b) What chemical is formed in the test for carbon dioxide?   (1)
(c) What do you understand by the terms (i) exothermic, (ii) slaking?   (2)

(d) Why is calcium carbonate added to the blast furnace in the manufacture of iron? (1)
(e) Suggest why farmers add calcium hydroxide to soil. (1)
(f) Name two industrial waste gases that could be removed by calcium hydroxide. (2)
(g) Suggest the raw material that could be used with limestone for the manufacture of sodium carbonate. (1)
(h) Name two chemicals that would react together to give calcium carbonate. Write the ionic equation for this reaction. (3)
(i) (i) Explain why calcium hydroxide removes temporary hardness but does not remove permanent hardness. Write the equation for the removal of temporary hardness. (3)
   (ii) What would happen to the water in the reservoirs if too much calcium hydroxide were added? (1)

16.6 Manufacture of sulphuric acid

Sulphuric acid is manufactured by the contact process. Sulphur dioxide is made by burning sulphur in air. The sulphur dioxide is mixed with air and passed over a catalyst of vanadium(V) oxide at 450°C. The sulphur dioxide reacts with the air to form sulphur trioxide; any unreacted sulphur dioxide is recycled.

The sulphur trioxide is dissolved in 98% sulphuric acid in a steel container. Water is added at the same time to keep the concentration of sulphuric acid at 98%.

Sulphur trioxide can be made in the laboratory using the apparatus shown below. The sulphur trioxide is dissolved in water to make sulphuric acid.

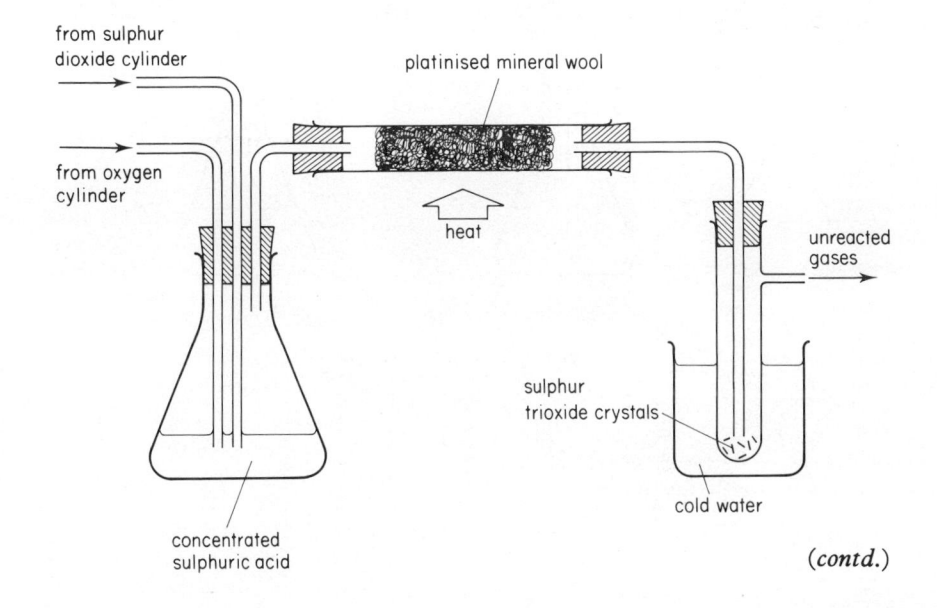

(*contd.*)

(a) Give six differences in the manufacture of sulphur trioxide between the laboratory method and the industrial method. (6)
(b) Give two large-scale uses of sulphuric acid. (2)
(c) A platinum catalyst used to be used in the contact process, but this catalyst was easily poisoned by arsenic.
   (i) What do you understand by the phrase 'the catalyst was easily poisoned'? (1)
   (ii) Substances that poison catalysts are also poisonous to living things. What name is given to proteins which catalyse reactions in living things? (1)
(d) Explain why the sulphur trioxide made by the contact process is mixed with nitrogen, oxygen and sulphur dioxide. (2)
(e) Why is the laboratory method of making sulphur trioxide performed in a fume cupboard? (1)
(f) Suggest an explanation for the following observations:
Increasing the temperature increases the speed of the reaction between sulphur dioxide and oxygen but decreases the yield of sulphur trioxide. (2)
(g) What was the percentage yield in a manufacturing process if 1.0 kg of sulphur trioxide was obtained from 5 kg of sulphur dioxide? (4)

16.7 Cigarettes

A cigarette was burnt in air using the apparatus shown in the diagram below.

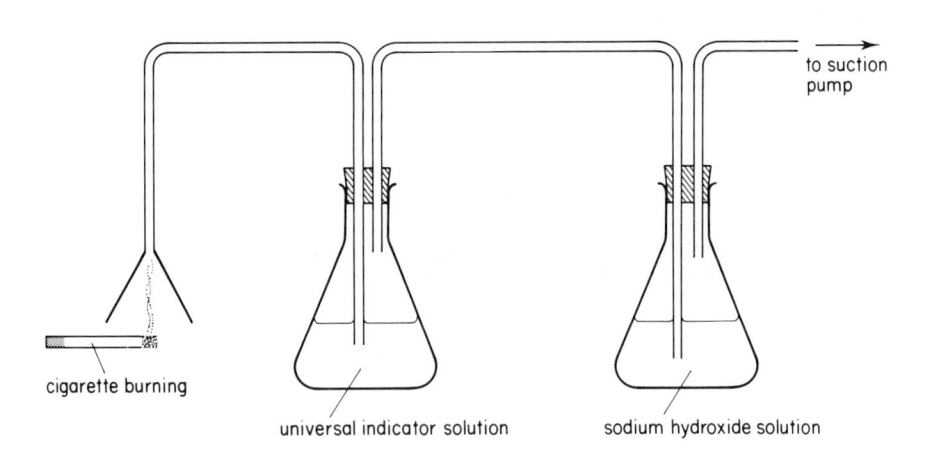

The pH of the solution formed was 5.0. When the products from burning a cigarette were investigated they were found to contain water vapour, carbon dioxide, carbon monoxide, nitrogen dioxide, sulphur dioxide and a chemical that causes cancer, 3, 4-benzopyrene ($C_{20}H_{12}$)

(a) Which four elements *must* be present in a cigarette? (2)
(b) Which of the products formed when a cigarette burns are acidic? (1)
(c) Which one of the gases formed when a cigarette burns will be coming out of the tube at X? Explain your answer. (2)
(d) What is the empirical formula of 3,4-benzopyrene? (1)
(e) How would you alter the apparatus in the diagram above to collect the water formed? You should only draw the apparatus for collecting water, and state where you would place it in the diagram. (3)
(f) Why is the suction pump necessary? (1)
(g) Briefly explain the toxic effect of carbon monoxide. (3)
(h) The figures below show the effect of smoking on health.

Type of smoker	% of CO breathed	% of carboxy-haemoglobin in blood	effect on health
non-smoker	0.0004	1.3	no effect
light smoker non-inhaler	0.001	2.3	acts on central nervous system
light smoker inhaler	0.002	3.8	acts on central nervous system
heavy smoker	0.010	25.0	breathing problems, headache, fatigue, heart problems, lung cancer

Suggest a reason why:
(i) there is carbon monoxide in the air breathed in by non-smokers (2)
(ii) heavy smoking causes breathing problems (2)
(iii) smoking causes lung cancer (1)
(iv) Predict the percentage of carboxyhaemoglobin in the blood of a very heavy smoker (0.020 % of carbon monoxide breathed in) (1)

17 Analysis

17.1 Tests for ions

List A gives a number of common ions. List B gives some common reagents.

List A   **List B**
chloride   sodium hydroxide solution
sulphate   dilute hydrochloric acid
nitrate   barium chloride solution
lead(II)   silver nitrate solution
copper(II)   dilute nitric acid
iron(III)   Devarda's alloy
carbonate

For each of the ions in List A, state
    (i) which of the reagents in List B are used to test for the ion
    (ii) what observation is made when the test is carried out.
No equations are required.   (4, 3, 3, 3, 2, 2, 3)

17.2 Analysis

Each of the substances A to E is an aqueous solution of one of the following: ammonium chloride, copper(II) sulphate, iron(II) iodide, lead(II) nitrate, sodium carbonate.
(a) Identify A to E from the following information, giving your reasons:
    (i) A gives a pale blue precipitate with dilute ammonia solution. The precipitate reacts with excess ammonia solution to form a dark blue solution.   (3)
    (ii) B gives no precipitate with sodium hydroxide solution, but when the solution is heated an alkaline gas is given off.   (3)
    (iii) C reacts with silver nitrate solution to form a yellow precipitate.   (3)
    (iv) D reacts rapidly with dilute hydrochloric acid to give a gas that turns lime water milky.   (3)
    (v) E reacts with sodium hydroxide to give a white precipitate. The white precipitate reacts with excess sodium hydroxide to form a colourless solution.   (3)
(b) Each of the reactions in **(a)** is a test for one of the ions in each of the substances A to E. Suggest tests by which you could identify the other ion present in each substance.   (10)

17.3 Analysis of a black compound

The diagram below shows some reactions that can be carried out starting with a black compound A.

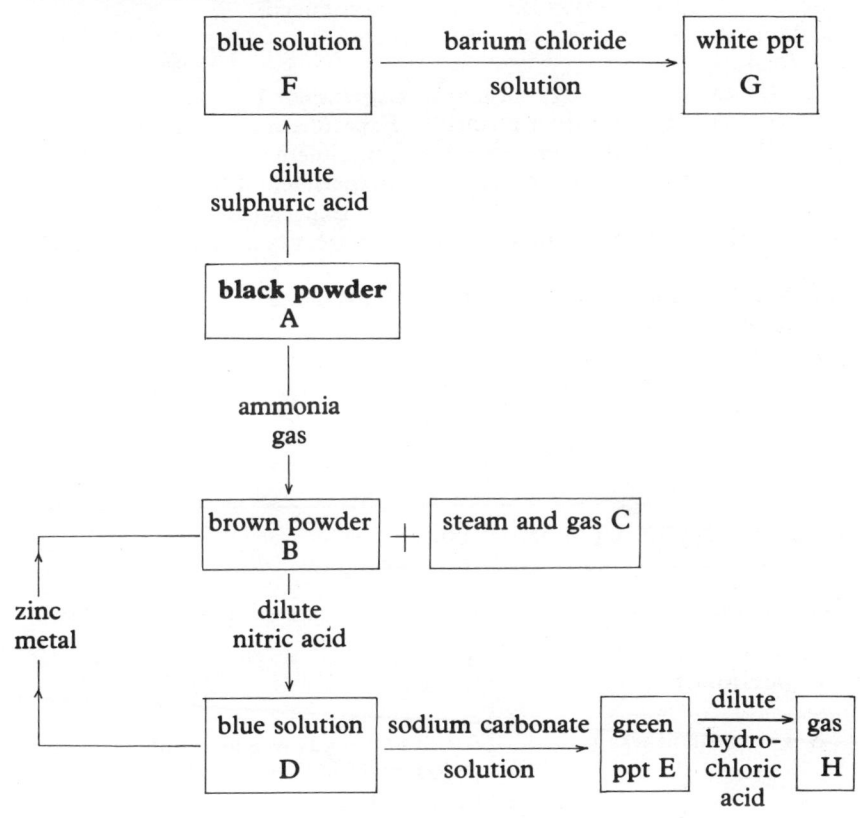

```
                    blue solution   barium chloride    white ppt
                         F          ─────────────→        G
                                       solution
                          ↑
                       dilute
                    sulphuric acid
                          │
                    ┌─────────────┐
                    │ black powder│
                    │      A      │
                    └─────────────┘
                          │
                       ammonia
                         gas
                          ↓
                    ┌─────────────┐   ┌──────────────────┐
                    │brown powder │ + │ steam and gas C  │
                    │      B      │   └──────────────────┘
                    └─────────────┘
         ↑                │
      zinc             dilute
      metal          nitric acid
         ↑                ↓
                                                       dilute
                    blue solution  sodium carbonate  green  hydro-   gas
                         D         ─────────────→   ppt E  chloric    H
                                       solution             acid
```

(a) Name each of the substances A to H. (8)
(b) (i) Write a molecular equation for the change from A to B. (1)
    (ii) Write the ionic equation for the change from D to E. (1)
(c) What use do we make in the laboratory of the chemical barium chloride used above in the conversion of F to G? (1)
(d) Briefly describe a method by which substance B could be made directly into substance A. (2)

17.4 Analysis involving iron compounds

Five experiments were carried out using iron and its compounds. The results obtained are described below.

Experiment 1: Iron was heated in steam. A black solid and hydrogen were produced.

Experiment 2: Iron was reacted with hydrochloric acid. A green solution and hydrogen were produced. *(contd.)*

Experiment 3: Some of the black solid from Experiment 1 was heated very strongly with aluminium powder. Iron and a white powder were produced.
Experiment 4: Chlorine gas was passed into the solution obtained in Experiment 2. A brown solution was produced.
Experiment 5: Some of the solution from Experiment 4 was treated with zinc. A colourless solution and iron were produced.

(a) Describe how you would test for hydrogen. (2)
(b) Identify
   (i) the black powder formed in Experiment 1 (1)
   (ii) the green solution formed in Experiment 2 (1)
   (iii) the white powder formed in Experiment 3 (1)
   (iv) the brown solution formed in Experiment 4 (1)
   (v) the colourless solution formed in Experiment 5. (1)
(c) Describe briefly how you would obtain pure, dry iron from Experiment 6. (3).
(d) Write equations for the reactions taking place in Experiment 1 and Experiment 4. (2)
(e) Suggest the products formed at (i) the anode (positive electrode) and (ii) the cathode (negative electrode), when the solution from Experiment 2 is electrolysed. (2)

17.5 Analysis of a mineral

The experiments described below were done on a white powder mineral.

Experiment	Result
A The mineral was heated.	A colourless gas was given off, leaving a yellow residue. The gas turned lime water milky.
B Excess dilute nitric acid was added to the mineral.	The same colourless gas as in **A** was evolved and a colourless solution obtained.
C The mineral was treated with excess hot, dilute hydrochloric acid.	After giving off the same gas as in **A** and **B**, a colourless solution was obtained, but white needle-like crystals formed on cooling the solution.
D The crystals from **C** were melted and electrolysed.	During electrolysis, a greenish-yellow gas was evolved at the anode (positive electrode) and a silvery molten ball of metal was formed at the bottom of the molten substance. It was found that 193 000 coulombs of electricity were required to deposit 1 mole of the metal.

(a) **(i)** Name the gas produced in Experiment **D**. How would you confirm the presence of this gas? (3)
  (ii) Explain why, in Experiment **D**, the metal does not form a coating on the cathode, but collects at the bottom of the melt. (2)
  (iii) What is the charge on the metal ion? (1)
(b) State three elements which must be present in the mineral and suggest an identity for the mineral. Give reasons for your answers. (6)
(c) Write down the formulae of all the ions present in the final solution obtained in Experiment **B**. (3)
(d) What can you say about the solubility of the needle-like crystals in Experiment **C**? (2)

18 Various topics combined

18.1 Uses of various substances

Give one everyday use for each of the following, other than in the laboratory.
- **(a)** ammonium sulphate (1)
- **(b)** calcium hydroxide (1)
- **(c)** chlorine gas (1)
- **(d)** copper metal (1)
- **(e)** a mixture of copper, nickel and zinc (1)
- **(f)** ethanoic acid (1)
- **(g)** ethanol (1)
- **(h)** helium (1)
- **(i)** mercury metal (1)
- **(j)** methane (1)
- **(k)** octane (1)
- **(l)** sodium chlorate(I) (1)
- **(m)** sodium chloride (1)
- **(n)** sucrose (1)
- **(o)** sulphur dioxide (1)

18.2 Tests for various gases

(a) Describe the action on damp pH paper of each of the following gases: ammonia   chlorine   hydrogen chloride   oxygen   (5)

(b) For each gas in turn, describe one other test which you could use to confirm its identity. (5)

18.3 Preparation of gases

The apparatus shown in the diagram below was used in separate experiments to make different gases.

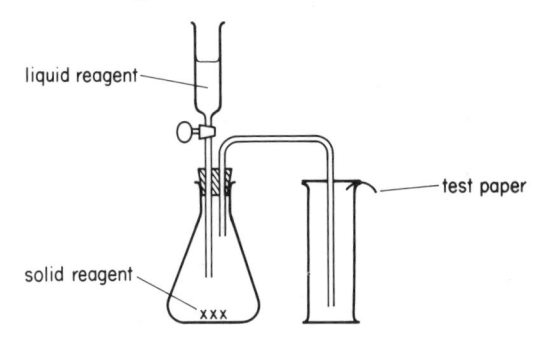

(a) Which two of following gases would you prepare and collect using the apparatus shown in the diagram?
ammonia   chlorine   hydrogen   hydrogen chloride   oxygen   (2)
(b) (i) For each gas you have chosen, say what reagents you would use. (4)
(ii) Write the equation for the preparation of each gas. (2)
(iii) For each gas, say what test paper you would use to find out when the gas jar is full, and what you would expect to happen to the paper. (4)
(c) Draw a diagram to show how you would modify the apparatus to produce a pure, dry sample of the gas. (4)

18.4 Reactions of sodium compounds

You are given four sodium salts.
A is sodium chloride solid.
B is anhydrous sodium sulphate.
C is hydrated sodium sulphite.
D is hydrated sodium carbonate.
(a) What ions are present in a dilute aqueous solution of B? (3)
(b) What gas is given off when dilute hydrochloric acid is added to C? How would you test for the presence of this gas? (3)
(c) What products are given off at the anode (positive electrode) and cathode (negative electrode) respectively when
(i) a concentrated aqueous solution of A is electrolysed (2)
(ii) a dilute aqueous solution of A is electrolysed? (1)
(d) Which one of these sodium compounds would not react with concentrated sulphuric acid? (1)
(e) The gas evolved by adding dilute hydrochloric acid to D is passed into a solution of calcium hydroxide until there is no more visible change. State all that is observed. (2)
(f) D has the formula Na_2CO_3, xH_2O. When 14.3 g of D was heated until no more water of crystallisation was given off, 5.3 g of the anhydrous salt remained. Calculate the value of x. (3)

18.5 False/true statements

Explain why each of the following statements is false. Your answers should include a diagram for (a) and (b) and an equation for (c).
(a) A covalent bond is formed when electrons are transferred from one atom to another. (3)
(b) A crystal of sodium chloride consists of molecules of sodium chloride arranged in a regular pattern. (4)
(c) The synthesis of water from its elements can be represented by the equation

$$H^+(aq) + OH^-(aq) \rightleftharpoons H_2O(l)$$ (3)   (*contd.*)

(d) Any solution of hydrogen chloride is a good conductor of electricity. (3)
(e) 1 mole of atoms of any gas occupies 24 dm³ at room temperature and pressure. (3)

18.6 Properties of elements

The following data is for the six elements A to F. Study this data, then answer the questions.

Element	m.p /°C	b.p /°C	Conducts electricity at room temperature	Volume that contains 1 mole of atoms under room conditions /cm³	Solubility in cold water
A	659	2470	yes	10.0	insoluble
B	−101	−34	no	12 000	soluble
C	−39	357	yes	14.8	insoluble
D	Sublimes above 3700		yes	5.4	insoluble
E	−249	−246	no	24 000	insoluble
F	−7	58	no	25.6	soluble

(a) Which element is a metal strong enough for a building material?
(b) Which element is a noble gas?
(c) Which element is mercury?
(d) What can you conclude about B and E from the difference in their volume per mole?
(e) Suggest the identity of the element labelled F.
(f) Suggest, with reasons, two elements from this list that might be in the same Group of the Periodic Table.
(g) Which element(s) would be liquid at (i) 100°C, (ii) −100°C? (10)

Data page 1

The Periodic Table

In each box:
- a = relative atomic mass
- X = atomic symbol
- b = atomic number

I	II	III	IV	V	VI	VII	VIII
1 H 1							4 He 2
7 Li 3	9 Be 4	11 B 5	12 C 6	14 N 7	16 O 8	19 F 9	20 Ne 10
23 Na 11	24 Mg 12	27 Al 13	28 Si 14	31 P 15	32 S 16	35.5 Cl 17	40 Ar 18
39 K 19	40 Ca 20	70 Ga 31	73 Ge 32	75 As 33	79 Se 34	80 Br 35	84 Kr 36
85 Rb 37	88 Sr 38	115 In 49	119 Sn 50	122 Sb 51	128 Te 52	127 I 53	131 Xe 54
133 Cs 55	137 Ba 56	204 Tl 81	207 Pb 82	209 Bi 83	— Po 84	— At 85	— Rn 86
— Fr 87	— Ra 88						

Transition elements (Period 4): 45 Sc 21, 48 Ti 22, 51 V 23, 52 Cr 24, 55 Mn 25, 56 Fe 26, 59 Co 27, 59 Ni 28, 64 Cu 29, 65 Zn 30

Transition elements (Period 5): 89 Y 39, 91 Zr 40, 93 Nb 41, 96 Mo 42, — Tc 43, 101 Ru 44, 103 Rh 45, 106 Pd 46, 108 Ag 47, 112 Cd 48

Transition elements (Period 6): La to Lu, 178 Hf 72, 181 Ta 73, 184 W 74, 186 Re 75, 190 Os 76, 192 Ir 77, 195 Pt 78, 197 Au 79, 201 Hg 80

Actinides: — Ac 89, — Th 90, — Pa 91, — U 92

Data page 2

1 mole of gas occupies 24 000 cm³ at room temperature and pressure.

The Faraday constant = 96 500 C (coulombs)/mol.

The Avogadro constant = 6×10^{23}/mol.

1 dm³ = 1000 cm³

In the questions and answers, the following alternative names could have been used:

chlorate(I)	or hypochlorite	ethanoate	or acetate
dichromate(VI)	or dichromate	ethanoic acid	or acetic acid
manganate(VII)	or permanganate	ethanol	or ethyl alcohol
nitrate	or nitrate(V)	ethene	or ethylene
nitric acid	or nitric(V) acid	methylbenzene	or toluene
sulphate	or sulphate(VI)		
sulphite	or sulphate(IV)		
sulphur(VI) oxide	or sulphur trioxide		
sulphuric acid	or sulphuric(VI) acid		